T0181453

Emer de Vattel and the Politics of Good Government

Antonio Trampus

Emer de Vattel and the Politics of Good Government

Constitutionalism, Small States
and the International System

palgrave
macmillan

Antonio Trampus
Linguistics and Comparative Cultural Studies
Ca' Foscari University Venice
Venice, Italy

ISBN 978-3-030-48026-4   ISBN 978-3-030-48024-0   (eBook)
https://doi.org/10.1007/978-3-030-48024-0

This Palgrave Macmillan imprint is published by the registered company Springer Nature Switzerland AG.
The registered company address is: Gewerbestrasse 11, 6330 Cham, Switzerland

ACKNOWLEDGEMENTS

The idea for this book originated in discussions with my friends and colleagues Koen Stapelbroek, Richard Whatmore, Béla Kapossy and Christoph Good during our visit to Vadstena in 2008 for the research conference *Reforming the European State System in the Long Eighteenth Century*, organised by the European Science Foundation. In the course of the subsequent ten years, I have pursued my research in many European archives, in Venice, Siena, Modena, Palermo, Bastia, Bern, Vienna, Paris and The Hague. I have been able to discuss the development of the phases of the work on several occasions with Antonella Alimento, Koen Stapelbroek, Gabriella Silvestrini and Walter Rech, during seminars hosted by the University of Pisa, the Erasmus University of Rotterdam, the Ca' Foscari University in Venice, and the Helsinki Centre for Intellectual History. More recently, some results of this research have been discussed at the University of Vercelli (Eastern Piedmont) within a section of the project *Natural Law 1625–1850* directed from Erfurt and Halle by Knud Haakonssen, Frank Grunert and Diethelm Klipper and coordinated in Italy by Gabriella Silvestrini and Elisabetta Fiocchi Malaspina. This research was supported in large part by the Italian Ministry of Education, University and Research (Research Project of Relevant National Interest 2015 "The Legacy of Enlightenment: Rights and Constitutionalism between Revolutions and Restorations 1789-1848" directed by Vincenzo Ferrone, University of Turin). The lengthy preparation and discussion of this project has initiated other new research by international scholars, opening new fields of investigation: some results of this have been published in the book *The Legacy of Vattel's Droit des gens* (Palgrave Macmillan, 2019), which I edited alongside Koen Stapelbroek and which in fact constitutes an important complement to this book. Special thanks to Matthew Armistead for his careful revision of the English text.

Much of this work has become intertwined over time with the research, reflections and contributions of Koen Stapelbroek, from whom I received inspiration, ideas and even study materials. My thanks go to him above all.

CONTENTS

CHAPTER 1

Introduction: The Invention of Good Government for the Law of Nations

For more than two hundred years Emer de Vattel's *Droit des gens* has attracted the attention of historians, jurists and political philosophers. The uninterrupted discussion and success of this work have been accompanied by lively debates about the originality and relevance of its author's ideas about crucial issues such as the position of the individual in international law, the right of war, the question of peace, republicanism, and the nature of the international order. The *Droit des gens*, originally published in Neuchâtel in 1758, has proved to be a text capable of crossing historical contexts and geographical space, in so doing becoming a bestseller of international law.

During this long journey through time, Vattel's treatise has, however, also been subjected to many processes that from a historical and interpretative point of view have distorted its meaning, processes that reflect the transformation of cultural contexts and changes in international geopolitics. The most significant is that after being used in the eighteenth century as a political and philosophical text invaluable to reform programmes and constitutional development, in the nineteenth century, after the Congress of Vienna, it began instead to be used as a simple yet dominant manual for studies on international law. Another important consequence of these processes has been the fact that compared to the period in which the *Droit des gens* played a part in political debates within states, in the past century interest in Vattel has turned almost completely towards his contribution to the analysis of interstate relations.

© The Author(s) 2020   1
A. Trampus, *Emer de Vattel and the Politics of Good Government*,
https://doi.org/10.1007/978-3-030-48024-0_1

However, the readers of the eighteenth and early nineteenth centuries perceived many more points of interest in the *Droit des gens*, and these aroused heated debates, attracted praise and criticism, and generated continuous re-editions, translations and adaptations of the text. This book offers a historical analysis that aims to show the deeper reasons for the success of Vattel's work, to understand what contemporaries saw and found in it and how it was able to make a contribution to the transformation of states and society, in Europe and across the Atlantic, both before and after the French Revolution.

To achieve this result, it is necessary to begin by shifting attention from the traditional vantage points from which Vattel's work has been considered.[1] The *Droit des gens* has in fact for a long time been analysed through internal readings of the text, in other words through philosophical or philological analyses aimed at studying the internal coherence of Vattel's thought in relation to the great past masters of natural law, and in so doing identifying his sources or his expositive and interpretative strategy. This, however, is only one of the possible approaches to the *Droit des gens*, and one that does not fully explain the reasons for the work's success, or the strategies for reading and receiving the text. If we turn our attention instead from the history of philosophy and law towards cultural and intellectual history, it is possible to study the legacy of the *Droit des gens* through the centuries, in particular in the second half of the eighteenth century and in the nineteenth, including from the perspective of the circulation of thought. This book also proposes a subsequent goal, namely that of turning attention further towards the strategies of appropriation of the text, and of the cultural 'consumption' of the *Droit des gens*, to understand not only how it was read, but also how it was used, adapted and reworked by those drafting constitutions, formulating reform projects and fighting everyday battles for freedom in different geographical, linguistic and social contexts within *ancien régime* Europe.

Also hidden behind these phenomena is the paradox of the asymmetrical relationship between the fame of the author and the way his work found success for reasons unconnected to his objectives. Furthermore, Vattel himself appealed on the one hand to an idea of a natural law that was universal and eternal, and which for this very reason was also the source of guarantees, and on the other hand was convinced that the great works of the previous century were no longer sufficient to explain the great transformation of commerce and society of his time. The result was a work that did its utmost to stick to principles and characteristics of generality and abstractness, with very few references to contemporary affairs so as not to risk becoming a victim of contingent situations and debates. This characteristic, which could have been a strength but also a weakness of the *Droit des gens*, was, however, destined to be quickly overtaken by events.

As documented by letters exchanged with the French censors, Vattel had tenaciously defended his work plan, which he developed by interweaving general principles with case studies carefully selected from history. Yet, almost immediately the tendency to use and bend the text to serve specific political and geographical situations became apparent. Even the early pirate edition published in Leyden in the same year as the first official one, 1758, revealed a readiness to insert in the text examples considered dangerous by the censors and by Vattel himself, which dragged the interpretation and use of the *Droit des gens* towards the specific interests of certain states.[2] To a greater or lesser extent, as we shall see in the pages of this book, this trend has continued to manifest itself constantly right up to the present day, through the strategies of translation and appropriation of the text, through the reorganisation of its contents in accordance with the tastes of editors, through summaries and epitomes used for study or work purposes in governmental offices, and through critiques and proposals for the correction and reconstruction of the work. The paradox of the *Droit des gens*'s success is linked precisely to these two opposing characteristics: on the one hand, Vattel's readers could find in it concepts, affirmations and general principles of natural law that were mostly well-known and therefore reassuring, and on the other, thanks to the way they were formulated, they could be immediately transferred into their own political contexts, which were those of nations in ever-increasing competition with each other.

GOOD GOVERNMENT: A RENAISSANCE PARADIGM FOR THE EIGHTEENTH-CENTURY STATE SYSTEM

All this happened, however, when Vattel could no longer communicate with his readers, because death had taken him in 1767, less than ten years after the publication of the work. The story of the *Droit des gens* thereupon became—as often happens in cultural history—also the story of a work without an author, in other words one of readings and interpretations beyond his intentions. Consequently, it is particularly important to try to understand what the historical situations were that made the use of the *Droit des gens* constantly relevant during the eighteenth and nineteenth centuries.

To this end it is necessary to return to the context in which Vattel wrote his work, that of mid-eighteenth-century Europe. As is recalled in the first chapter, the 1713 Peace of Utrecht and the 1718 Peace of Passarowitz had profoundly changed the international landscape, both through religious pacification and through the recognition that competition between states was by then more and more frequently of a commercial nature, rather than purely military. In consequence of this, the ancient moral code of conduct between states was gradually replaced by a system of legal norms based on the political will of the princes and nations. It was therefore necessary to look for the theoretical bases of this new system, and the books of the past no longer provided satisfactory answers, nor did they use a language suited to the new times.

Vattel was aware of these changes and above all of the fact that to understand the new international order it was necessary to start from the point of departure, to understand the nature and function of the state, and to use natural law to understand the mechanisms through which political relationships between single individuals and between the various communities of people are formed. He dedicated the first part of his work to this research, giving a long and detailed explanation of what exactly was meant by states, nations and constitutions, and also what was meant by sovereignty. The first book, in short, appeared more as a treatise on constitutional philosophy, which explained what the minimum criteria should be for achieving formal equality between states and which, in this sense, laid the foundations for modern constitutionalism.[3] Only in the second part of the work did Vattel focus on themes that today are considered more typical of international law, in other words relations between states, the international system, treaties and their interpretation, and the politics of alliances and the remedies to apply when these failed.

As much of this research illustrates, public attention in the eighteenth century was primarily directed towards the first part of the *Droit des gens*. This is surprising, and it begs the question of what people found so interesting in those pages or, rather, why Vattel's investigation into the constitution of the states seemed so important. The second chapter of this book focuses on the fact that running through Vattel's work is a repeated reference to a phrase corresponding to a key concept of the modern age, which is that of 'good government,' understood as the ability of those who govern to adopt effective laws to avoid clashes and build consensus within the community. By using this concept, which became a key point in his plan for a constitutional state and for the regulation of international relations, Vattel managed to bridge the great cultural, philosophical and political debate of modern Europe on the goodness of the laws and the qualities of a good legislator, the foundations of which went back to antiquity and had returned to the centre of attention with the crisis of the *ancien régime*.

In eighteenth-century Europe, the idea and the need for a good government—construed as a constitutional practice capable of acting as regulator and mediator not only within the sphere of politics but also in that of commerce, serving as a remedy for the new inequalities that this produced—became an increasingly relevant issue. The myth of the good government, suggested by the intellectual legacy of the ancient world and its revival in the world of the Renaissance,[4] was invoked as a regulating principle and a restraining influence on the power politics of the great monarchies, in the face of absolute sovereigns trying to be perceived as good legislators. It therefore offered a formula of guarantee for a constantly changing reality coming to terms with the great political and economic transformations of the crisis of the *ancien régime*.

This debate also relaunched the role of the small republican states. In these states, according to the classical and Renaissance tradition, virtue and political moderation depended on the laws and not on an individual and absolute legislator, while awareness and care for the wellbeing of the population were a direct consequence of the absence of particular interests and of economic interests that were subject to power politics.

Vattel's focus on the good government thus became, in the interpretations of his contemporaries, a strong argument in favour of republicanism and the essential survival of the small states. At the same time, the pages which the *Droit des gens* dedicated to the internal constitution of states, to the subject of political consensus and the legitimacy of sovereignty, indicated that to reach formal equality between the states it was necessary to study the mechanisms through which the political will of a nation was expressed, the problem of the legitimisation of sovereignty, and, finally, the origins and the function of constituent power in line with a concept central in modern constitutionalism that Vattel, in 1758, was one of the first to define in this way. As we also see in this chapter, this was an argument of interest mainly to the small states of central and southern Europe, from the German states to Switzerland and the Italian peninsula, where ancient political and cultural traditions relating to good government that could act as a moderating influence on the princes already existed, and where these were based on principles of peace and justice that came to be identified with Ambrogio Lorenzetti's famous Medieval fresco cycle, *The Allegory of Good and Bad Government*, which still decorates Siena's Palazzo Pubblico.[5] In this way, the language of Vattel seemed on the one hand to be very modern, while on the other it referenced older traditions and realised them in the light of the problems and political changes of the eighteenth century.

The first case for the reception of the *Droit des gens* in terms of these modes of interpretation was that of the Mediterranean islands, which towards the middle of the eighteenth century found themselves at the centre of the manoeuvring between the great powers (Spain, Austria and France) and attempted to assert their traditions of autonomy and resistance when threatened by governments that were often geographically and culturally distant. Vattel's work, with its idea that an essential characteristic of the good government and of the constitutional state had to be the willingness to make the interests of the nations the central consideration, thus provided unexpected ideological support.

As we see in the third chapter, this was the case of Sicily, which after the Peace of Utrecht (1713) passed from the dominion of Spain to the control of Victor Amadeus II of Savoy, and then to Austria (1718). Soon after, this was also the case of Corsica, which in 1728 rebelled against the Republic of Genoa to try to win independence, which it in fact briefly achieved until it ceded to France in 1768. On these two Mediterranean islands, whose affairs had significant effects on the political and constitutional debate in Britain and America, Vattel's work was read, studied and commented on immediately after its publication in 1758, both to justify the right to resist foreign governments that attacked ancient political freedoms, and to understand how to interpret more favourably the international treaties that impacted on the destinies of those territories.

VATTEL THE 'CONSTITUTIONALIST'

At the start of the 1780s the political function of the *Droit des gens* would become even clearer in the eyes, this time, of a prince ruling over a state which, though small, had great cultural influence over the entire European continent. Peter Leopold of Habsburg was the second son of Francis Stephen, who had governed Lorraine independently of France up to the

1736 exchange with the Grand Duchy of Tuscany, before marrying the future empress Maria Theresa of Austria. Like his father, Peter Leopold undertook an extensive programme of economic, religious and administrative reforms that accelerated when news of the War of Independence began to arrive from the American colonies. Convinced of the possibility of transforming Tuscany into a constitutional monarchy, Peter Leopold instigated, with his advisers, a series of in-depth studies of the French, English and American political cultures, and of Vattel's work, with the aim of bringing to fruition a constitutional project capable of interpreting the philosophy of the good government in the light of new ideas of political liberty and of the Tuscan tradition of citizen's freedoms. The fourth chapter of this volume shows, by analysing the different phases in the writing of Peter Leopold's constitution, how the *Droit des gens*—a philosophical and theoretical work—was recast as an aid to the drafting of the constitution, offering concepts, ideas and key words that were transferred almost directly into Peter Leopold's manuscripts.

The use of Vattel's work in the constitutional experiment of Tuscany did not, however, exhaust its function in the debate on the position of the small state, but rather contributed to restoring the importance of the *Droit des gens* in the international context of the 1770s. Indeed, while in Holland and Switzerland new editions which announced the discovery of previously unpublished additions by Vattel were being prepared, in Tuscany work began on an Italian translation limited to the first volume dedicated to the constitutional structure of the state and the concept of the nation. This was a difficult task complicated by the fact that it involved transferring and adapting from French into Italian as well as from the language of the 1760s to that of the 1780s—a decade influenced by the spirit of the Atlantic Revolutions—words and concepts that were taking on new meanings. The research carried out for this volume in the State Archives of Siena, the city of the *Good Government* frescoes, has led to the rediscovery of the lost manuscript of this first Italian translation, and the various stages involved in this translation, along with the choices made by the translator, and the impact that these had on the political debate both in Tuscany and on the nearby Kingdom of Naples are reconstructed in the fifth chapter.

By the time that the Tuscan translation was completed, however, the echo of the *Droit des gens*, and the various readings and interpretations of it on both sides of the Atlantic Ocean, in Europe and in the American colonies had already transformed the work into a political and literary sensation. The *Droit des gens* no longer served only to legitimise the existence of small states or the right of resistance of the inhabitants of Corsica and of the American colonies, but also to satisfy the appetite of the book market and public opinion. This was recognised by a journalist working in Venice, Lodovico Antonio Loschi, a friend of the celebrated adventurer Ange Goudar, the author of *The Chinese Spy* (1764). Loschi prided himself on being aware of all the latest literary developments, and in no time at all had prepared and published another Italian translation of the *Droit des gens* that immediately captured the hearts of the public and rendered the printing of the Tuscan edition useless. Chapter 6 explains that the Venetian version of Vattel's work, printed in 1781, reflected not only the international political problems of the time, but also the tastes of the public and the interests of the governing authorities of Venice, who wanted to use the *Droit des gens* as a theoretical instrument in addressing the decline of the republic and in promoting the latest efforts of political reform.

The End of an International Utopia

The situation in Venice, which in the midst of its political and economic decline found itself involved in a difficult international controversy with Holland that risked triggering armed conflict, allows us to see how use was made of the *Droit des gens* in late eighteenth-century diplomacy and commercial competition. As Chap. 7 demonstrates, in the last two decades of that century, Vattel's work had already become a classic of international politics, an authoritative source from which to derive principles and rules of help in analysing and solving problems and crises in the internal politics of states and of the European and the Atlantic world. The chapter tells the history of a commercial fraud which, starting in 1771, saw Holland, Venice and Austria set against each other, with their respective ambassadors all victims of adventurers and swindlers who brought the three countries to the brink of a European war. Vattel's work then became the point of reference for those who, resorting to international arbitration, saw the good government of the European balance of powers—through the use of reason rather than military action—as the key to resolving the conflict and safeguarding the prestige and dignity of the two republics, which were also small states that could easily fall prey to international power politics.

In more or less the same years, as we see in Chap. 8, when facing the progressive affirmation of the language of rights through the *American*

Declaration of Independence and then through the French *Declaration of the Rights of Man and of the Citizen*, the *Droit des gens* was used and discussed as a synthesis of the natural law tradition and therefore as a useful point of reference for the debate on the natural rights of man, both by the culture of the Enlightenment and that of Catholicism, especially in the Mediterranean region. Catholic thinkers in particular formulated a complex strategy of exegesis around the *Droit des gens*, enabling them to recognise the importance of Vattel's work while making use of it for anti-Protestant purposes, to criticise the foundation of the right to resistance (one of the most important human rights in the era of Atlantic revolutions) and to emphasise instead other rights and duties such as those of service, assistance and asylum that were linked to the virtues of charity and humanity and were of help in reinforcing the natural bonds within society.

For over thirty years after its publication, the *Droit des gens* had thus been the work through which it was possible to measure how the meaning of such terms as 'state,' 'constitution,' 'nation' and 'republic' changed when having to deal with international power politics, trade conflicts and the crisis of the *ancien régime*. It can therefore be understood why, during the years of the French Revolution and the Napoleonic Empire, Vattel's text was viewed with suspicion and even as a danger. Depending on the situation and the interpretations made of it, it could be used as a 'revolutionary' work to emphasise the right of resistance against an oppressor, the role of the nation before the state or within the state, and the role of the small states as a constraint and antidote to the despotism of the great powers and empires.

Chapter 9 of this volume shows how, after the era of Napoleonic hegemony in Europe, the *Droit des gens* reacquired its importance in the preparations for the Congress of Vienna and the geopolitical redrawing of nineteenth-century Europe. The idea that the small states, whether monarchies or republics, could continue to play a part in the new balance between the powers led to further success, a 'renaissance' of Vattel's work. At the same time, however, the book was also deprived of its typically eighteenth-century utopian and political charge in favour of a community of free and equal constitutional states. In short, Vattel's work was neutralised, with its political function set aside in order to become an essentially technical manual of international law.

At the beginning of the nineteenth century and especially after the Restoration, the profound changes in the meaning of words like 'homeland,' 'nation' and 'constitution' made it clear that it was no longer possible to read the *Droit des gens* with eighteenth-century eyes. To be sure, the parts of the work dealing with international relations between the states could still be useful in the new contexts of nineteenth-century diplomacy, but the first book, on the subject of the internal constitution of the states, had to be radically reinterpreted in order to avoid offering arguments that might prove dangerous to supporters of liberal constitutions and democratic revolutions. The European culture of the Restoration thus set in motion a multifaceted operation aimed at divesting the *Droit des gens* of its nature as a philosophical and political work, and thereby of the potential political project that was central to that nature, in order to transform it into a simple university textbook of international law.

At the same time, European scholars and commentators, mostly in Germany, Italy and Portugal, initiated a radical critique of the first book of the *Droit des gens*, which in some cases actually involved revising and reworking the text. As is shown in the twelfth and final chapter of this book, there remained some interpreters who still tried to use Vattel's work in a subversive sense, in defence of the freedoms and rights of nations and individuals, as happened in the political trials of mid-nineteenth-century Italy. Others attempted to use it to call into question the idea that the positive laws of the state could prevail over the natural law of communities and over individual safeguards. But among the majority of the interpreters there prevailed instead the idea that the *Droit des gens* was a historical document of a bygone age, a text that was no longer relevant, and a source that was simply technical, which presented basic concepts of international law but was not enough to elucidate with any effectiveness political and international reality. This 'renaissance' of the *Droit des gens*—its second life as a manual for use in diplomatic life—was, however, destined to produce another divergence between European culture and the culture of the United States and Latin America. There, as explained in the conclusion, Vattel's work would experience another rebirth, continuing to play a role of stimulus and point of discussion in the public and political debate.

Given the nature of this volume, which engages with the reception, diffusion and translation of ideas through different editions of the *Droit des gens* published in different languages, certain choices had to be made regarding the citation and presentation of these various editions. Throughout the book, references will take the same standardised form and include the book, chapter and paragraph of the *Droit des gens*, so as to help readers using different standard editions of the text in different languages. Most of the quotations come from the widely available

English-language edition of 2008, which includes an introduction by Béla Kapossy and Richard Whatmore (Indianapolis: Liberty Fund) and maintains the text and English title *The Law of Nations: Or, Principles of the Law of Nature Applied to the Conduct and Affairs of Nations and Sovereigns* of the 1797 London standard edition. References to this English-language edition, however, still take the same form as above, citing Vattel, *Droit des gens*, book and chapter in Roman numerals, and paragraph in Arabic numerals. In cases where differences in language, concepts or contexts are relevant and are part of the argument being made, references and quotations are from other editions or unpublished translation manuscripts, as specified.

NOTES

1. For a comprehensive bibliography on Vattel see Andrew Hurrell, "Vattel: Pluralism and its Limits," in *Classical Theories of International Relations*, ed. Ian Clark and Iver Neumann (Houndmills-New York: Palgrave, 2001), 233–255; Béla Kapossy and Richard Wathmore, "Introduction," to *Law of Nations: Or, Principles of the Law of Nature, Applied to The Conduct and Affairs of Nations and Sovereigns*, ed. Béla Kapossy and Richard Whatmore (Indianapolis: Liberty Fund 2008); *Vattel's International Law in a XXIst Century Perspective/Le droit International de Vattel vu du XXIe siècle*, ed. Vincent Chetail and Peter Haggenmacher (The Hague: Martinus Nijhoff, 2011).

2. Édouard Beguelin, "En souvenir de Vattel (1714–1767)," in *Extrait du Recueil des travaux offerts par la Faculté de droit de l'Université de Neuchâtel à la Société suisse des Juristes à l'occasion de sa réunion à Neuchâtel 15–17 septembre 1929* (Neuchâtel: Université de Neuchâtel, 1929), 93; Antonella Alimento, "The French Reception of Vattel's Droit des gens: Politics and Publishing Strategies," in *The Legacy of Vattel's Droit des gens*, ed. Koen Stapelbroek and Antonio Trampus (Cham: Palgrave Macmillan, 2019), 135–164.

3. In this sense one can speak of Emer de Vattel's 'constitutionalism' and I am in agreement with the considerations expressed by Martti Koskenniemi, "'International Community' from Dante to Vattel," in Chetail and Haggenmacher *Vattel's International Law/Le droit International de Vattel*, 51–75.

4. See Quentin Skinner, *The Foundation of Modern Political Thought*, vol. 1, *The Renaissance* (Cambridge: Cambridge University Press, 1978), especially part two.

5. The invitation to explore the implications of the idea of good government in the commercial policies of the eighteenth century has recently also been issued by Sophus A. Reinert, *The Academy of Fisticuffs: Political Economy and Commercial Society in Enlightenment Italy* (Cambridge Mass.: Harvard University Press, 2019).

BIBLIOGRAPHY

Most of the quotations from Vattel's *Droit des gens* come from the widely available English-language edition of 2008, which includes an introduction by Béla Kapossy and Richard Whatmore (Indianapolis: Liberty Fund) and maintains the text and English title *The Law of Nations: Or, Principles of the Law of Nature Applied to the Conduct and Affairs of Nations and Sovereigns* of the 1797 London standard edition.

Contemporary translations of primary sources are listed under the names of their authors. Full manuscript sources are referenced only in the notes for reasons of space.

PRIMARY SOURCES

PRINTED BOOKS

Vattel, Emer de, *Law of Nations: Or, Principles of the Law of Nature, Applied to The Conduct and Affairs of Nations and Sovereigns*, eds. Béla Kapossy and Richard Whatmore (Indianapolis: Liberty Fund 2008)

SECONDARY SOURCES

Alimento, Antonella, "The French Reception of Vattel's Droit des gens: Politics and Publishing Strategies," in *The Legacy of Vattel's Droit des gens*, eds. Koen Stapelbroek and Antonio Trampus (Cham: Palgrave Macmillan, 2019), 135–164

Beguelin, Édouard, "En souvenir de Vattel (1714–1767)," in *Extrait du Recueil des travaux offerts par la Faculté de droit de l'Université de Neuchâtel à la Société suisse des Juristes à l'occasion de sa réunion à Neuchâtel 15–17 septembre 1929* (Neuchâtel: Université de Neuchâtel, 1929)

Chetail, Vincent and Haggenmacher, Peter, eds., *Vattel's International Law in a XXIst Century Perspective/Le droit International de Vattel vu du XXIe siècle* (The Hague: Martinus Nijhoff, 2011)

Hurrell, Andrew, *Vattel: Pluralism and its Limits*, in *Classical Theories of International Relations*, eds. Ian Clark and Iver Neumann (Houndmills-New York: Palgrave, 2001), 233–255

Kapossy, Béla, and Wathmore, Richard, "Introduction," to *Law of Nations: Or, Principles of the Law of Nature, Applied to The Conduct and Affairs of Nations and Sovereigns,* eds. Béla Kapossy and Richard Whatmore (Indianapolis: Liberty Fund 2008), ix–xx

Koskenniemi, Martii, ""International Community" from Dante to Vattel," in *Vattel's International Law in a XXIst Century Perspective/Le droit International de Vattel vu du XXIe siècle,* eds. Vincent Chetail and Peter Haggenmacher (The Hague: Martinus Nijhoff, 2011), 51–75

Reinert, Sophus A., *The Academy of Fisticuffs: Political Economy and Commercial Society in Enlightenment Italy* (Cambridge Mass.: Harvard University Press, 2019)

Skinner, Quentin, *The Foundation of Modern Political Thought,* vol. 1, *The Renaissance* (Cambridge: Cambridge University Press, 1978)

CHAPTER 2

Vattel's *Droit des gens*. A Transnational Bestseller from the Age of Enlightenment

The *Droit des gens* is fundamental to understanding the evolution of the political debate that arose during the Seven Years' War on the system of international relations and the relationship between the interests of states of trade, and the discussion relating to natural law that was had during the crisis of the *ancien régime*. Interpreters, be they Vattel's contemporaries or more recent ones, have used the work in very different ways, sometimes regarding it as an effective reinterpretation of the tradition of natural law, and at other times as a republican argumentation against despotism.[1] What is certain, however, is that Vattel offered a useful redefinition of natural law that brought the traditions associated with Hugo Grotius and Christian Wolff into dialogue with the exigencies of the law of nations, which laid the groundwork for a modern doctrine of the fundamental rights of individuals and states.[2]

Because of these very qualities, Vattel's work enjoyed widespread success, as is shown by the numerous editions and translations: twenty in France, ten in Britain, a dozen in the United States, as well as others in Italy, Spain and Latin America.[3]

The critical literature, often of high quality, has, however, spotlighted the theoretical dimension and the internal analysis of the *Droit des gens*, leaving some other important questions in the shade. The first relates to

---

Archival Abbreviations
ASV   Archivio di Stato di Venezia, Venice, Italy

© The Author(s) 2020   15
A. Trampus, *Emer de Vattel and the Politics of Good Government*,
https://doi.org/10.1007/978-3-030-48024-0_2

the different phases of the reworking and consolidation of the book. This long exercise began with the first edition, which was published during the Seven Years' War, and continued with the versions of the 1770s, rewritten even as the independence of the American colonies was being debated. It finally drew to a close with the early nineteenth-century editions, during the pre- and post-Restoration period of international political balances. By contrast, the second question relates to how individual states, politicians and men of letters used Vattel's work to good effect in their political and institutional reform programmes and, also, to how it impacted on foreign policy procedure and international relations.

The Peace of Utrecht (1713) as the Premise for a New Europe

From a theoretical point of view, the Peace of Utrecht is the indispensable premise for any understanding of the importance of Vattel's work. From a historical perspective it was also the most important testbed for a new way of constructing international politics and of regulating relationships between states during the eighteenth century. The Utrecht agreement did not end in a single treaty, nor, in a technical sense, did it immediately achieve the goals that a peace process normally pursues in the context of international relations. Rather, it signalled the start of a new practice of international and interstate relations that was destined to produce long-term effects. Alongside the three so-called multilateral agreements—of Utrecht, Rastatt and Basel—another twenty bilateral accords were signed following lengthy negotiations. It should be remembered that at the time of the Utrecht treaties in the European political lexicon the term 'peace' still retained multiple meanings that have subsequently been lost as the word was reduced to narrower semantic fields typical of our own time: in that period 'peace' not only signified the opposite of war, and therefore disarmament or armed vigilance, but it also expressed concepts of security, conservation and tranquillity in the domestic sphere, that is, of the sphere within the state, and in the international public sphere.[4] In the political language of the eighteenth century, 'security,' 'conservation' and 'tranquillity' were all terms that were used much more frequently than the word 'peace.'

The debate on the significance and the real effects of the Utrecht agreements on the international politics of the eighteenth century is still unsettled. Some now consider them, as some did at the time, as a means of consolidating the dynasties in the form of political equilibrium, others instead as the origins of a collective security system.[5] The agreements developed around the Dutch city certainly gave rise to a watershed in the history of interstate relations. The Utrecht agreements helped develop diplomacy as a political and social institution, and it was in them, in particular in the first part of the treaty between Spain and England, that the expression 'balance of power' first began to be used purposefully. The concept, which was mentioned for the first time in 1701 by Charles Davenant in his *Essays on the Balance of Power* and reappeared with the same meaning in the 19 April 1709 issue of Daniel Defoe's publication *A Review of the Affairs of France*, thus came into general usage in the talks concerning the war and the conferences that led to the signing of the Treaty of Utrecht.

As regards Europe, with Utrecht the centre of gravity of international relations shifted significantly towards the Mediterranean. Attention focused on Spain and the rejection of its control over the Italian peninsula. Spanish dominion had often been preferred by the Italians because it was exercised by a more distant, blander, less centralised and less efficient government,[6] whereas the Habsburgs, taking the place of the Bourbons, established a new dynasty in Naples itself. It is true that the Kingdom of Naples passed from one form of dependency to another, but it is equally the case that this was the moment, with the creation of a 'national' monarchy,[7] when a new chapter opened in the history of the Mediterranean linked to new forms of political organisation of communities, to new areas of autonomy and to the search for forms of economic and political freedom.

This freedom was construed, to some extent throughout Europe, mainly in terms of the independence and autonomy of the small states from the influence of the great powers. The survival strategies of the small states reopened ancient discussions on the political virtue inspired by the size of these countries, and raised new questions about the importance of neutrality as a guarantor of independence. European culture harboured a long tradition of exalting the virtues of the small states, which were immune to the interests and politics of power and so tended to be averse to any form of despotism.[8]

As for neutrality, even during the War of the Spanish Succession, Venice, Rome and Florence had discussed the possibility of guaranteeing their continuation through the creation of a league of small states.[9] In Piedmont,

Victor Amadeus, the head of a small state which, thanks to an alliance with England, had obtained an advantageous peace settlement from France, had, from the end of 1712, suggested to the Queen of England that the Emperor's use of the English fleet to transfer the imperial troops from Spain to Italy be made conditional on the neutrality of the Italian peninsula. Then, in January 1713, the neutrality of the Italian princes was also requested by France as a guarantee against the Habsburgs, and this was supported by Venice. According to the political language of the time, neutrality would serve the "security" of the Duke of Savoy, the "safest repose" of the peninsula, and the containment of the Empire.[10]

A reflection of these issues is provided by a famous report by the Venetian ambassador Carlo Ruzzini, the permanent representative to the States General at The Hague and a special envoy during the Utrecht negotiations. Ruzzini had been very clear, during and after the negotiations, that the main change in the interstate system was not so much the shifting of the dynastic axis from Spain to Austria. Rather, he had understood that the real protagonists of the international scene were once again the small states, as demonstrated by the role that the Duke of Savoy had assumed with the support of anti-French and anti-Empire England. He thought it was a situation that might present an opportunity for other players, like Venice. He therefore noted that the Emperor and the Empire at the centre of Europe felt pressing "on their heart new, stronger forces and sharper punctures," and thus suffered "with disdain the initiative of the Duke of Savoy and the support lent by England and France."[11]

The concessions obtained with English help by the Duke of Savoy had, according to Ruzzini, surprised everyone. And they had the after-effect of the duke now being able to deny foreign powers access to the Italian peninsula. As long as he "does not voluntarily open the doors, it seems that France cannot find any way to put a foot in Italy."[12] From the standpoint of the Republic of Venice, this was a highly desirable situation since it and the Duchy of Savoy—two small states, hitherto minor players on the international stage—were left to watch over the security of the peninsula on their opposite borders, to the west and east, and thus found themselves allies. But Ruzzini feared that this apparent balance could vanish and that the Duke of Savoy would bring new upheaval to Italy, because of his dynastic ties, because of the political influence exercised by Eugene of Savoy at the Habsburg court, and also because Marie Adelaide, mother of the future Louis XV, though not yet destined for the throne, was also a Savoy.

The stakes, then, were obviously high: not only was the containment of the European powers in question, but so too was the ability of the small states to erect a geopolitical and strategic barrier against the large ones. Ruzzini, however, made clear that the problem of the small states and their security was not just an Italian or Germanic issue, but rather a pan-European one, for he mentioned the importance of "Spanish Flanders," that is, the Austrian Netherlands, as a rampart against France and the United Provinces. The Dutch, Ruzzini believed, preferred the proximity of the emperor, resident in Flanders, to that of the elector of Bavaria, because in the former they had a neighbouring prince who was better able to defend them against the French king.[13] Another case at the centre of international attention was that of Switzerland, not only because some of the peace negotiations were taking place there, but also because the confederation of cantons was another place in which to test the role of the small states and the system of balances. It should not be forgotten that precisely as a result of Utrecht and the new European settlement the great powers had sanctioned the passing of Neuchâtel—the home of Jean de Barbeyrac, one of the future protagonists of the European debate on natural law and the law of nations—into the hands of the Hohenzollern.[14]

The small states thus acquired an awareness of their role on the European stage and sought to assert it more in the diplomatic sphere than the commercial and territorial ones, concentrating attention on the rules and rituals of diplomacy, and on protocol and ceremony, as the small Republic of Lucca and the Duke of Savoy insisted on doing.[15] The use of ceremony and the observation of formalities still seemed to be able to demonstrate that, irrespective of grandeur, it was possible to overcome the distinction between small and large states, between real power and relative power, to give the impression that all states were formally equal, having recourse to a 'culture of appearance' that served to carry forward the plan for a European balance.

The other standard by which the policies of the states were measured was that of neutrality. In the case of the Republic of Venice, as Ruzzini wrote, the decision to adopt this political approach did not mean that Venice "should remain excluded from the participation and the honour of these peace [accords]," but was in fact a mark of political independence and autonomy.[16]

Indeed, on the same day as the signing of the suspension of hostilities between the Duchy of Savoy and France, 14 March 1713, the convention between England and the Habsburg monarchy was also agreed to ensure the neutrality of the Italian peninsula and the limitation of imperial troops within it. The agreement between the great powers thus revitalised the myth of the small state, of its virtuous neutrality as a factor of European security. This myth was then fuelled by publications of the time by way of extensive literary and iconographic propaganda, and through a variety of works like that of Casimir Freschot, who in 1715 published *The compleat history of the Treaty of Utrecht*, a 600-page volume with a dedication to the two English delegates to the negotiations, or like that of Jean Le Clerc, the author of a largely apologetic history of the United Provinces.[17]

These happenings opened the way to many plans for universal peace and to a rich literary and utopian trend, starting with the work by the Abbot of Saint-Pierre, that would look upon the Peace of Utrecht as the birth of a new era of perpetual concord. In the realm of diplomacy, a more pragmatic phase began that aimed to bring about new rules for the resolution of international disputes through the elaboration of a law of nations that would take into consideration not only the needs of states and governments but also the aspirations of people, of nations and homelands, that is, the potential new actors in international politics that were important for creating a universal society and consolidating reciprocal agreements.[18] According to the vision of some of the protagonists of the time, the increased number of subjects in play would make it possible to keep in check the great powers and impose the principle of reciprocity as a balancing force in international relations, as prescribed by article 2 of the Treaty of Utrecht, which pointed to the *iustum Potentiae equilibrium* as the foundation of the peace and tranquillity of the Christian peoples.[19]

THE CONSEQUENCES OF RELIGIOUS PEACE: LEGAL IMPERATIVES TOWARDS MORAL IMPERATIVES

The Utrecht treaties thus represent an important change of direction for the theoretical development of international relations and, in particular, for the law of nations. The political and philosophical culture up to that point had based the interpretation of the law of nations principally on *De iure naturae et gentium* by Samuel von Pufendorf, who had elaborated the theoretical principle of inequality between states as a precondition for stable alliances. After Utrecht, attention instead turned mainly to Hugo Grotius and his *De jure belli ac bacis*, which was based on a different idea that was more compatible with the order being sought, that is, on the theory of the natural equality among all peoples, suitable for creating a *consensus gentium* necessary for the stability of the international system.

It was Jean Barbeyrac who mediated and reworked the positions of these two authors, combining the study of the law of nations with attention to the role and function of the small European states. Barbeyrac had studied in Geneva, and taught in Lausanne and then Groningen in the United Provinces. He had already made a name for himself as a translator and commentator of Pufendorf's works: in 1706 he translated *De iure naturae et gentium* and in 1707 *De officio hominis et civis*, following these with his own original works on ethics and history, such as *Traté du jeu* (1709), *Traité de la Morale des Pères de l'Église* (1728) and *Histoire des Anciens Traités* (1739).

In the wake of the Utrecht treaties, of the new European order and of the role played by the small states, Barbeyrac increasingly shifted his focus from the work of Pufendorf to that of Grotius, which he translated in 1724 with the intention of establishing an organic system of the law of nations that would enable a practical use of both authors' conceptions of natural law within the renewed international setting. Barbeyrac knew that the Utrecht peace treaty had, in its innovative way, attributed greater legitimacy in the system of international relations to the concept of 'nation' alongside the notion of 'state.' From the second half of the seventeenth century the term 'state' (*État*) had already become part of the language of international relations, as had the term *puissance*, but with different meanings, not to refer to an international actor necessarily leaning towards a foreign policy based on power, but rather to explain the characteristics of the sovereign state, that is, the requirement of national sovereignty as a prerequisite for recognition by other states and legitimacy to act on a juridical-international level.

One of Barbeyrac's first objectives was to translate Grotius from Latin to French, a project that he brought to conclusion in 1724. Hence in the years following the Peace of Utrecht, Grotius's ideas spread internationally through this French version, primarily in the small European states, where it was accompanied by intense commentary work and translations into other 'national' languages. Barbeyrac believed that it was necessary to create an organic system of the law of nations that connected Pufendorf and Grotius, both Protestant authors, but his project was even more ambitious than this, and included both a new organisation of the law of nations and the possibility of building bridges with the Catholic culture. As his interpreters have shown, he used a model of translation that brought together the theological premises linked to the need to disseminate natural law doctrines with methodological declarations on the centrality of natural law as the foundation of the law of nations. He thus helped to shift the spotlight from natural law as a purely philosophical system towards natural law as the source of the law of nations. At the same time, he introduced what has been defined as a displacement operation which moved legal thought away from the Protestant world to the Catholic one, making the two authors' works accessible to Catholic culture. His translation of and commentary on Grotius allowed him to use the law of nations to set out a symmetrical system that united the study of moral philosophy, so dear to Catholicism, with the investigation of constitutional systems (in other words, the internal stability of states) and the international system (i.e. the stability of the system of states). Thanks to this mediation, the thought of Grotius, after Utrecht, became a more secure guide to accompany in Italy "the often tormented and restless paths of moral philosophy."[20] The reception of Grotius involved questions of a juridical, political and also ethical nature, and through him the idea of a more reasonable form of Christianity that clashed with the intransigent and conservative dogmatism of some sectors of Catholic culture was disseminated. In this way the Dutch author could also be used, in the name of a form of natural law that would be more rational and easier to share, to oppose the Enlightenment interpretations of natural law, which is to say against those who supported a naturalistic and utilitarian idea of natural law.[21]

Grotius's work from there on spread more widely in the Mediterranean area. In doing so it introduced into the political culture of the small Italian states the principle of the *consensus gentium*, that is, the consensus of all peoples or the consensus of the most civilised peoples for the formation of an international society, and also that of the *appetitus societatis*. Both these principles were criticised by Pufendorf, who preferred a utilitarian *socialitas*. This theoretical discontinuity in the continuity of natural law was underlined by many eighteenth-century authors, for example in 1751 by Giuseppe Maria Buondelmonti in an essay that would be read out publicly in Florence's Accademia della Crusca under the title *Ragionamento sul diritto della guerra giusta*.

It was in this context that Emer de Vattel's *Droit des gens*, published in 1758, intervened to construct a systematising and complex line of reasoning that used natural law to link the problem of security and stability of the international order to that of the internal constitutional stability of each state. To do so he concentrated not so much on the small states as territorial entities, as on nations as political organisations of the people. His reasoning started from the concept of natural equality between men, their rights and their duties originating from natural law, and then was extended to nations seeing that they were composed of men and perceiving themselves as free people who lived together in a state of nature and therefore shared the same rights and obligations.[22]

To formulate this theory Vattel had to consider the entire tradition of European natural law. Grotius, in his opinion, had ultimately reduced the law of nations to the simple uses of the nations, to construct instead a voluntary law of nations dependent on the *consensus gentium*; in consequence, he had underestimated the fact that, in any event, political societies and nations lived in a state of reciprocal interdependency.[23] Moving on, Vattel argued that Pufendorf presented certain limits in so far as he had maintained the confusion between natural law and the law of nations without trying to draw a clear distinction between them. The attempt to mediate between the two authors might have been helped by the writings of another great theorist of natural law, Christian Wolff, which Vattel had edited and commented on and which constituted an important point of reference for his own ideas. In particular, Vattel called attention to two brief essays on natural law written by Wolff: the *Essai sur le fondement du droit naturel et sur le premier principe de l'obligation où se trouvent les hommes d'en observer les lois* and the *Dissertation sur cette question: La loi naturelle peut-elle porter la société à la perfection sans le secours des loi politiques?*

The first of these was a sort of analysis of the fundamental concepts of natural law, primarily those of the law and of morality. It was "a general theory of the duties of man, regarded simply as man, or a science which teaches us *what is naturally good or bad* for man, *what he must do* and what *he should not do*" (my italics). Vattel noted that in the common language these expressions were often confused "because they all have the same purpose, namely to regulate the customs and conduct of men." By distinguishing the legal rule from the moral rule it was, however, possible to understand the difference between what was right and good in itself and what instead was right because it conformed to the duties of man and society. But all this was not enough, Vattel went on, because people not only needed to understand the importance of these laws, but also needed to observe them. This turned the spotlight on the concept of 'obligation,' which for Vattel meant being bound or, more accurately, being almost morally driven to do something despite being a free individual, and, moreover, to do so even against the instinct of passions and without being forced or physically constrained. The study of natural law helped one to understand these phenomena and to grasp the fundamentals of the functioning of civil and interstate society since, to quote Cicero, natural law is "ratio summa insita in natura, quae iubet ea quae facienda sunt, prohibet que contraria." Unlike natural laws, civil laws were norms of behaviour established by the civil authority. They could not oppose natural law since "at heart they coincide with it" but they made up for the "shortcomings" of natural law when dealing with the too frequent wrong actions of men.

According to many interpreters, this theoretical construction implied the subordination of ethics to politics, and, according to the most recent scholars, derived from Vattel's "non-juristic concept of rights and obligations."[24] For Vattel, then, the laws of nature were not perfect, and natural law itself, like morality, was inherently incomplete. Only through voluntary law, as a positive law, does natural law become complete and can be considered a real law accompanied by the faculty of obligation and constraint.

This reasoning may seem complex and segmented but it was well understood by eighteenth-century European readers and intellectuals. It appeared convincing first because it conveyed an idea of morality no longer anchored to the theological dimension but closely correlated to natural law. Second, it explained that morality, being considered a pre-juridical and pre-political concept, was a force capable of orienting the course of law, of society and of the nations but in itself incomplete without its activation through voluntary law. Moreover, the study of morality helped to qualify duties as a necessary corollary for activating rights. The close connection between moral imperatives, construed above all as man's duties towards society, and legal imperatives construed as cardinal obligations of the theory of human rights would remain a central theme in eighteenth-century constitutionalism.

MORALITY AND TRADE AFTER PASSAROWITZ

It is not possible to fully grasp the context in which the *Droit des gens* was born without considering the history of the treaty and the peace which, five years after Utrecht, modified the balances of Eastern and Mediterranean Europe. Still today the Treaty of Passarowitz (1718) features prominently in the analyses of historians because it seems to have been the first situation in the history of the modern age in which a model of negotiation emerged that replaced the competition of arms with that of trade. The name of Passarowitz, a distant place now known as Požarevac, in Serbia, was destined within a few years to open the eyes of Europeans to the new problem of commercial rivalries, to economic competition, to the "jealousy of trade" that Hume described as the key characteristic of a modernity based on the development of free trade even in its most radical consequences.[25] In short, in the history of international relations the Peace of Passarowitz provides important early evidence of the awareness that the regulation of commercial relationships could serve as an effective antidote to war. Until then commercial clauses remained isolated within more general peace treaties, but after Passarowitz the idea gradually emerged that commercial treaties should be separated from peace treaties, even if the two types of treaty were signed at the same time.[26] In a second phase, the purpose of commercial treaties became not only that of ending wars, but also that of modifying economic arrangements and forms of competition in order to preclude war in itself.[27]

When we examine the Passarowitz case in more detail, we find that this symmetry is confirmed by the fact that the peace treaty between the emperor and Ahmed III, and the Republic of Venice and the Ottoman sultan (21 July 1718) was followed by a new trade and navigation treaty between the Habsburg monarchy and the Ottoman Empire (27 July 1718). As the most recent historiography has observed, the division of material between the two treaties implied that the peace agreement implicitly took into account its impact on the function of war itself, in the sense that it established free exchange rights and trade rules independently of the outcome of the war.[28]

We see, then, that the Peace of Passarowitz was a corollary of that of Utrecht, and read together the two made clear the transition from the logic of arms to that of trade, especially when it was necessary to choose the most suitable means for regulating or restoring order to the international context. Starting from Passarowitz, therefore, it became increasingly evident that trade treaties could be instruments used to curb the politics of power and to obviate war or mitigate its effects. Hume had this picture in mind when he described the tendency of modern politics to self-destruct and to annihilate the principles of civilisation on which it was built,[29] and Vattel himself shared this idea, taking his place in the process that John Pocock called the "Utrecht Enlightenment."[30]

These changes also directly involved the Mediterranean area in which the upshot of Utrecht and Passarowitz could be gauged and the new political and institutional solutions to problems like those posed by Hume and Vattel could be tested. The question of commercial and military sovereignty over the Mediterranean and its divisions (particularly the Adriatic, Ionian and Aegean Seas) should not therefore be considered, with regard to the eighteenth century, as a simple peripheral reflection of an eastern conflict with the Ottoman Empire. Rather it was a testbed for a new form of power politics and for new interstate European formations.

Passarowitz is important in the picture we have outlined also for another reason: it was actually the treaty that succeeded in removing for the first time one of the classic causes of justification of war, namely that based on religious diversity or, more specifically, on the need to defend Christianity when it was endangered or intimidated. In the course of the eighteenth century the new phenomenon in the Mediterranean would therefore become that of wars waged not with arms but by means of trade. The extension and amplification of commercial competition demonstrated, on the one hand, that military operations were ever more often combined or concealed by commercial operations and, on the other, that the use of armed force was no longer a necessary aspect of the prosecution of a war, and hence it became increasingly difficult to distinguish clearly between wartime and peacetime. Many wars, especially trade wars, would be fought while peace reigned between the contending sides.

In the Mediterranean these phenomena were augmented by the use by many states, and especially those with a southern coastline subject to the Ottoman Empire, of privateering as a means of harming their rivals economically. There was frequent recourse to the granting of special authorisations—permitted by custom and the law of nations—to carry out guerrilla actions in the absence of a formally declared war. The commercial shipping of distant countries such as Holland and Denmark paid the highest price, being the object of constant attacks by French and imperial corsairs. More and more often the problem became that of reconciling the use of privateering with the idea of free trade and the development of trade as a means of neutralising the use of war.[31] The rejection of armed warfare and the limitation of privateering were both necessary to counter the use of violence, which was incompatible with economic development and with the wellbeing of peoples, cities and ports.

How could these problems best be tackled? The seventeenth-century theories of both the law of nations and the classical doctrine on the times and ways of just war (derived mostly from Grotius and Wolff) failed to give effective answers predicated on natural law alone. As Vattel would note, classical natural law theory tended to link the problem and discussion of just or legal war only to relations between states, and, accordingly, left the discussion about the *ius ad bellum* exclusively to the princes. It was therefore necessary to reread natural law in the light of the changes being made in the mid-eighteenth century and of the emergence of these new subjects and international competitors, different as they were to states and sovereigns. Vattel also reflected on this point in the *Droit des gens*, which is another reason why his work received such a welcome and success.

Vattel: Natural Law as a Moral Code for States

In 1747 Vattel had begun to address these scenarios and, in particular in the context of the new dynamics of European conflicts, wrote a *Mémoire et autres pièces concernant la création et l'objet d'une raprésentation diplomatique de la Cour de Dresde à Berne*, in which he recommended that Prussia hand Neuchâtel over to Saxony or Poland on the grounds of the principle that Neuchâtel, as a free and sovereign principality, could legitimately undergo this transfer in furtherance of its interests and a better balance in the European area. The year before Vattel had entered the service of King August III of Poland, and had thereupon set out his ideas in the knowledge that he would find authoritative interlocutors, probably including his cousins, the Montmollins, who were members of Neuchâtel's *Conseil d'État*. His idea was presented in a memorandum to the King of Poland,[32] and starting from these thoughts and from the need to demonstrate that even a small state like Neuchâtel could legitimately express its will and sovereignty in the international context, he began his reflection on the role of states not as abstract territorial entities, but as products of the political organisation of peoples and nations. Consequently, the different level of power effectively created between states could never refute the principle of this natural inequality. To use Vattel's own analogy, a dwarf could very well look like a giant and a small republic was no less a sovereign state than a powerful kingdom.

Christian Wolff had already formed a theory on the "freedom of the states" (*libertas civitatis*), regarded as the basis for explaining what the notion of state interest consisted of with respect to obligations imposed by treaties. Vattel deemed the work useful but distanced himself from Wolff's belief that a *civitas maxima* could actually exist, that is, that there could be a system of cogent rules that bound nations according to natural law, thereby reducing the incidence of voluntary law.[33]

It was therefore necessary, according to Vattel, to begin to reflect on the relationship between natural law and the voluntary law of nations and, consequently, on the subjects who produced this law. It was no longer sufficient to look only to the state or the government of the prince, but instead it was imperative to bring dignity to the nations as societies of men rooted in a territory and connected to each other by reciprocal rights and duties. In this way Vattel outlined a new participant in international relations, one capable of moving beyond the classic concept of the state and the politics of power that was based mainly on the criterion of territorial extension. The result was that Vattel's "small" state was no longer Jean Bodin's "petit roy"[34] but could obtain greater prestige and become a virtuous example capable of preventing both domestic and therefore internal despotism as well as the desire for conquest by empires, and could, in other words, become a balancing factor on the international stage.[35]

In this way he renewed in depth the thought on the dynastic, economic and military interests that—through Pufendorf—had guided the politics of power in the modern age and placed the importance of the interests of society and its wellbeing in the foreground.[36] In an age in which the debate on the purpose and usefulness of republicanism was reopened, the *Droit des gens* helped to make the case for the tradition of the small states. Vattel's position was even more important if one considers the conflicting opinion of Bielefeld, author of the *Institutions politiques* (1760) and advisor to Frederick II of Prussia, who believed the small states to be completely anachronistic in a world of major new powers. For him the only hope for their survival consisted in their perfect neutrality, the moderation and wisdom of their governors, and the politics of self-preservation; otherwise they would be part of a utopian world and be forced to choose, for safety's sake, to abandon neutrality and non-alignment and to lean on the strongest.

It is clear, then, that the *Droit des gens* was a valuable tool to employ when seeking to justify the continued existence of the small states and, more generally, for all European states which did not pursue power politics, preferring the status of neutrality. For that reason, Vattel's work was not perceived as being only a textbook or a simple synthesis of the European natural law tradition but assumed much greater importance.

In a long review of the *Droit des gens*, published in 1759 in the April and May issues of the *Journal de commerce*, it was noted that Vattel's natural law theory matched his economic ideas of "the true principles [...] of the most natural order."[37] Within this order Vattel attached a primary importance to agricultural development, not to modern commerce and luxury, which had since the later seventeenth century taken centre stage in political efforts to spur economic growth. States could rise to greatness, and all states together benefit from each other's greatness, in a commercially friendly political Balance of Power system if agriculture was generally accepted as the key to national economic growth. The *Journal de commerce* made much of the central role of agriculture in Vattel's *Droit des gens* and identified him as a potentially radical anti-Colbertist.[38]

Vattel was therefore very aware of the need for reform, as well as of the complications of this challenge. It would seem that he saw states behaving in the international realm in the same way as individuals who had been over-socialised and made sensitive to pride, dignity and honour. The real problem was the existence of unnatural competition between states, which had arisen because national economies had not developed and integrated with each other in the way they should have. The result was that belligerents used sovereign rights to interfere with the trade of other states. The same rights, he argued, now ought to be deployed to monitor the balance of power through the creation of a well-calibrated system of commercial treaties that shaped patterns of commercial exchange and the internal hierarchy and relative force of Europe's globalised national economies.[39] This was a major challenge. Mutual aid among individuals as well as states was a basic principle of natural law that Vattel had adopted from Wolff. But the issue of which economic policies could be used to make this moral principle into a reality and thereby reform the interstate system was a matter broadly debated in Europe. Rather than merely stipulating that natural law dictated the integration of national economies, since these were complementary and resulted from the cultivation of available natural resources, and then deriving any principles of free or protected trade from this stipulation, by declaring that the balance of power was the object of the voluntary law of nations Vattel made commerce subservient to the realisation of a political ideal, that is, to the stability of the balance of power. With regard to commercial treaties, states thus remained free to conclude treaties that in their view stabilised the balance of power and thereby promoted mutual aid among states.

Vattel's reluctance to interfere directly with the most prominent legal problems of neutrality and trade of the time had already earned him the reputation of a writer who failed to protect trade against the excesses of warfare. There was a specific way in which Vattel's *Droit des gens* chimed well with the predicament of the European small states in the later eighteenth century. While the political glory of the small states referred to the defunct model of the Renaissance city state and its principle of republican virtue, the *Droit des gens* offered an alternative political justification based on the concept of good government. This concept, to which Vattel devoted many pages of the *Droit des gens*, is central to understanding the similarities and differences between the classical tradition, natural law and the birth of political modernity expressed in both the democratic and liberal experience of the nineteenth century.

NOTES

1. Andrew Hurrell, "*Vattel: Pluralism and its Limits,*" in *Classical Theories of International Relations*, ed. Ian Clark and Iver Neumann (Houndmills-New York: Palgrave, 2001), 233–255; Nicholas Greenwood Onuf, "Civitas Maxima: Wolff, Vattel and the fate of Republicanism," *American Journal of International Law* 88 (1994): 287–296; David Boucher, *Political Theories of International Relations: From Thucydides to the Present* (Oxford: Oxford University Press, 1998), 255–268; Richard Tuck, *The Rights of War and Peace: Political Thought and the International Order From Grotius to Kant* (Oxford: Oxford University Press: 1999), 191–196; Albert Geouffre de Lapradelle, "Introduction" to Emer de Vattel, *Le Droit des Gens ou principes de la loi naturelle appliqués à la conduite et aux affaires des nations et des souverains* (Washington: Carnegie Institution, 1916), i–lv; Christoph Good, *Emer de Vattel (1714–1767). Naturrechtliche Ansätze einer Menschenrechtsidee und des humanitäre Völkerrecht im Zeitalter der Aufklärung* (Zurich: Dike Verlag AG, 2011).

2. Tuck, *The Rights of War and Peace*, 191–196; Isaac Nakhimovsky, "Vattel's Theory of the International Order: Commerce and Balance of Power in the Law of Nations," *History of European Ideas* 33 (2007): 157–173; Richard Whatmore and Béla Kapossy, "Emer de Vattel's Mélanges de littérature, de morale et de politique (1760)," *History of European Ideas* 34 (2008): 77–103.

3. A complete list of translations is offered by Elisabetta Fiocchi Malaspina, *L'eterno ritorno del Droit des gens di emer de Vattel (secc. XVIII–XIX). L'impatto sulla cultura giuridica in prospettiva globale* (Frankfurt: Max Planck Institute for Legal History, 2017), 262–272; a new edition of Vattel's first English translation (1760) is *Law of Nations: Or, Principles of the Law of Nature, Applied to The Conduct and Affairs of Nations and Sovereigns*, ed. Béla Kapossy and Richard Whatmore (Indianapolis: Liberty Fund 2008).

4. Inken Schmidt-Voges, "Making Peace in Early Modern Europe," in *Peace Was Made Here: The Treaties of Utrecht, Rastatt and Baden 1713–1714*, ed. R.E. Bruin and M. Brinkman (Petersberg: Im Hof, 2013), 51; see also Emmanuelle Jouannet, "Les dualismes du Droit des gens," in *Vattel's International Law from a XXIst Century Perspective*, ed. Vincent Chetail and Peter Haggenmacher (Leiden: Brill, 2011), 133–150.

5. David Onnekink, "The treaty of Utrecht 1713," in Bruin and Brinkmann, *Peace Was Made Here*, 66; Jean-Pierre Bois, *De la paix des rois à l'ordre des empereurs 1714–1815* (Paris: Editions du Seuil, 2003), 26.

6. Raffaele Aiello, *Preilluminismo giuridico e tentativi di codificazione nel regno di Napoli* (Naples: Jovene, 1965), 89; see also *Good government in Spanish Naples*, ed. Antonio Calabria and John A. Marino (Berlin-New York: Peter Lang, 1990).

7. Aurelio Musi, "La nazione napoletana prima della nazione italiana," in *Nazioni d'Italia. Identità politiche e appartenenze regionali tra Settecento e Ottocento*, ed. Angela De Benedictis, Irene Fosi and Luca Mannori (Rome: Viella, 2012), 84.

8. Matthias Schnettger, *"Kleinstaaten in der Frühen Neuzeit. Konturen eines Forschungsfeldes,"* Historische Zeitschrift cclxxxvi (2008): 605–640.

9. Annibale Bozzola, *Giudizi e previsioni della diplomazia medicea sulla Casa di Savoia durante la guerra di successione di Spagna* (Turin: G. Bonis e Rossi, 1914), 22–24.

10. Annibale Bozzola, "Venezia e Savoia al congresso di Utrecht (1712–1713)," *Bollettino storico-bibliografico subalpino* XXXV.3–4 (1933): 30–39.

11. Carlo Ruzzini, "Relatione del Congresso di Utrecht di miser Carlo Ruzzini, Kav. E Procurator, Ambasciatore estraordinario, plenipotenziario, 1713," in: *Venetiaantsche Berichten Berichten over de Vereenigde Nederlanden 1600–1795*, ed. Pieter J. Blok (Gravenhage: Martinus Nijhoff, 1909), 341–370; ASV, *Relazioni degli ambasciatori*, no. 15.

12. Blok, *Venetiaantsche Berichten*, 368.

13. Blok, *Venetiaantsche Berichten*, 367.

14. Louis Grandpierre, *Histoire du Canton de Neuchâtel sous le Roi de Prusse 1707–1848* (Leipzig: Grandpierre, 1805), 15 ff.; Louis-Edouard Roulet, "Friderich der Große und Neuenburg," in *Friedrich der Große in seiner Zeit*, ed. O. Hauser (Cologne-Vienna: Böhlau, 1987), 181; Frédéric-Alexandre de Chambrier, *Histoire de Neuchâtel et Valangin jusqu'à l'avènement de la Maison de Prusse* (Neuchâtel: Attinger, 1840), 494–509; Adrian Bachmann, *Die prussiche Sukzession in Neuchâtel-Ein ständisches Verfahren um die Landesherrschaft im Spannungsfeld zwischen Recht und Utilitarismus (1694–1715)* (Zurich: Schulthess, 1993), 193 ff.

15. Renzo Sabbatini and Paola Volpini, eds., *Sulla diplomazia in età moderna. Politica, economia, religione* (Milan: FrancoAngeli, 2011), 116–117.

16. Blok, *Venetiaantsche Berichten*, 368.
17. Jean Le Clerc, *Geschiedenissen der Vereenigde Nederlanden, sedert den aanvang van die Republyk tot op den Vrede van Utrecht in 't Jaar 1713* (Amsterdam: Zacharias Chatelain, 1738).
18. Vattel, *Droit des gens*, Preliminaries, book II, ch. 1 and books III and IV.
19. See Vincenzo Buonomo, "Reciprocità, libertà religiosa e protezione dei diritti umani in ambito internazionale," in *Libertà religiosa e reciprocità*, ed. Jose Antonio Arana Mesa (Milan: Giuffrè, 2009), 123–124.
20. Maurizio Bazzoli, *Il piccolo Stato nell'età moderna: Studi su un concetto della politica internazionale tra XVI e XVIII secolo* (Milan: Jaca Book, 1990), 56.
21. Bazzoli, *Il piccolo Stato nell'età moderna*, 57.
22. Bazzoli, *Il piccolo Stato nell'età moderna*, 61–62; Vattel, *Droit des gens*, Preliminaries, § 18.
23. Vattel, *Droit des gens*, Preface, ch. VIII.
24. Peter Paul Remec, *The Position of the Individual in International Law according to Grotius and Vattel* (The Hague: Martinus Nijhoff, 1960), 197.
25. Christine Lebeau, "Negotiating a Trade Treaty in the Imperial Context: The Habsburg Monarchy in the Eighteenth Century," in: *The Politics of Commercial Treaties in the Eighteenth Century. Balance of Power, Balance of Trade*, ed. Antonella Alimento and Koen Stapelbroek (Cham: Palgrave Macmillan, 2017), 365–369; Istvan Hont, *Jealousy of Trade: International Competition and the Nation-State in Historical Perspective* (Cambridge, MA: Harvard University Press, 2005).
26. Stephen C. Neff, "Peace and prosperity: commercial aspects of peacemaking," in *Peace treaties and International Law* in: *European History. From the Late Middle Ages to World War One*, ed. Randall Lesaffer (Cambridge: Cambridge University Press, 2004), 365–381.
27. Ana Crespo Solana, "Cooperation or Neutrality? How War Affects Business Strategies: The Case of Cadiz, Spain (1700–1720)," in *Neutres et neutralité dans l'espace atlantique durant le long XVIIIe siècle (1700–1820)*, ed. Éric Schnakenbourg (Bécherel: Les Perséides, 2015), 31–58.
28. Koen Stapelbroek, "'The long peace': commercial treaties and the principles of global trade at the Peace of Utrecht," in: *The 1713 Peace of Utrecht and its enduring effects*, ed. A.H.A. Soons (Leiden: Brill, 2018); Koen Stapelbroek, "L'organisation du commerce international dans l'ombre d'Utrecht: les perspectives hollandaises du XVIIIe siècle," in *La paix d'Utrecht (1713): Enjeux économiques, maritimes et commerciaux*, ed. Lucien Bély, Géraud Poumarède and Guillaume Hanotin (Paris: Pedone, 2018), 475–502.

29. David Hume, *Essays Moral, Political, and Literary*, ed. Eugene Miller (Indianapolis: Liberty Fund, 1985). For context, see Istvan Hont, *Jealousy of Trade*.

30. John G. A. Pocock, *Barbarism and Religion*, vol. 1, *The Enlightenments of Edward Gibbon, 1737–1764* (Cambridge: Cambridge University Press, 1999) and John G. A. Pocock, *Barbarism and Religion*, vol. 2, *Narratives of Civil Government* (Cambridge: Cambridge University Press, 1999).

31. Eric Schnakenbourg, *Entre la guerre et la paix. Neutralité et relations internationales XVIIe–XVIIIe siècles* (Rennes: Presses Universitaires de Rennes, 2013), 75–121.

32. This memorandum can be found transcribed in its entirety in Edouard Beguelin, "En souvenir de Vattel (1714–1767)," in *Extrait du Recueil des travaux offerts par la Faculté de droit de l'Université de Neuchâtel à la Société suisse des Juristes à l'occasion de sa réunion à Neuchâtel 15–17 septembre 1929* (Neuchâtel: Université de Neuchâtel, 1929)), 132–134. Tetsuya Toyoda, "Vattel's doctrine of national sovereignty in the context of Saxony Poland and Neuchâtel," in *Theory and politics on the law of nations: Political Bias in International Law Discourse of Seven German Court Councilors in the Seventeenth and Eighteenth Centuries* (Leiden: Brill, 2011), 161–190.

33. Theodore Christov, "Vattel's Rousseau: Jus Gentium and the Natural Liberty of States," in *Western Political Science Association 2011 Annual Meeting Paper*. Available at: https://ssrn.com/abstract=1766921, 167–169.

34. Jean Bodin, *Le six livres de la République* (Paris: Puys, 1583), book 1, ch. II, § 13; Douglas M. Johnston, *Historical Foundations of World Order: The Tower and the Arena* (Leiden-Boston: Martinus Nijhoff, 2008), 340–341; Ann Blair, "Authorial Strategies in Bodin," in *The Reception of Bodin*, ed. Howell A. Lloyd (Leiden-Boston: Brill, 2008) 137–156.

35. Maurizio Bazzoli, *Stagioni e teorie della società internazionale* (Milan: LED, 2005), 392; Emilio Gabba and Aldo Schiavone, eds., *Polis e piccolo Stato tra riflessione antica e pensiero moderno (atti della giornate di studio 21–22 febbraio 1997)* (Florence: New Press, 1997).

36. Nakhimovsky, *Vattel's Theory of the International Order*, 157–173.

37. *Journal de commerce* 1759, 141. Quoted by Koen Stapelbroek, "Universal Society, Commerce and the Rights of Neutral Trade: Martin Hübner, Emer de Vattel and Ferdinando Galiani," in: *Universalism in International Law and Political Philosophy*, ed. Petter Korkman and Virpi Mäkinen, *Collegium. Studies across Disciplines in the Humanities and Social Sciences* 4 (2008) (Helsinki: Helsinki Collegium for Advanced Studies, 2008), 63–89.

38. *Journal de commerce* 1759, 152–155; Koen Stapelbroek and Antonio Trampus, "The Legacy of Vattel's Droit des gens: Contexts, Concepts, Reception, Translation and Diffusion," in: *The Legacy of Vattel's Droit des*

gens, ed. Stapelbroek and Trampus, 12. Antonella Alimento, "Tra strategie editoriali e progettualità riformista: la circolazione in Francia de Le droit des gens di Emer de Vattel," *Rivista Storica Italiana* 129 (2017): 548–558.
39. Nakhimovsky, *Vattel's Theory of the International Order*, 157–173.

BIBLIOGRAPHY

Most of the quotations of Vattel's *Droit des gens* come from the widely available English-language edition of 2008, which includes an introduction by Béla Kapossy and Richard Whatmore (Indianapolis: Liberty Fund) and maintains the text and English title *The Law of Nations: Or, Principles of the Law of Nature Applied to the Conduct and Affairs of Nations and Sovereigns* of the 1797 London standard edition.

Contemporary translations of primary sources are listed under the names of their authors. Full manuscript sources are referenced only in the notes for reasons of space.

ARCHIVAL ABBREVIATIONS

ASV   Archivio di Stato di Venezia, Venice, Italy

PRIMARY SOURCES

JOURNALS

Journal de commerce, vols. 2–3, April–May (Brussels: Van den Berghen, 1759)

PRINTED BOOKS

Bodin, Jean, *Le six livres de la République* (Paris: Puys, 1583)
Hume, David, *Essays Moral, Political, and Literary*, ed. Eugene Miller (Indianapolis: Liberty Fund, 1985)
Le Clerc, Jean, *Geschiedenissen der Vereenigde Nederlanden, sedert den aanvang van die Republyk tot op den Vrede van Utrecht in 't Jaar 1713* (Amsterdam: Zacharias Chatelain, 1738)
Lichtenberg, Georg Christoph, *Briefwechsel*, eds. Ulrich Joost and Albrecht Schöne, 3 vols. (Munich: Beck, 1983–1985)
Ruzzini, Carlo, "Relatione del Congresso di Utrecht di miser Carlo Ruzzini, Kav. E Procurator, Ambasciatore estraordinario, plenipotenziario, 1713," in *Venetiaantsche Berichten over de Vereenigde Nederlanden 1600–1795*, ed. Pieter J. Blok (Gravenhage: Martinus Nijhoff, 1909)

Vattel, Emer de, *Law of Nations: Or, Principles of the Law of Nature, Applied to The Conduct and Affairs of Nations and Sovereigns*, eds. Béla Kapossy and Richard Whatmore (Indianapolis: Liberty Fund 2008)

SECONDARY SOURCES

Aiello, Raffele, *Preilluminismo giuridico e tentativi di codificazione nel regno di Napoli* (Naples: Jovene, 1968)
Alimento, Antonella, "Tra strategie editoriali e progettualità riformista: la circolazione in Francia de Le droit des gens di Emer de Vattel," *Rivista Storica Italiana* 129 (2017), 548–558
Bachmann, Adrian, *Die preussiche Sukzession in Neuchâtel-Ein ständisches Verfahren um die Landesherrschaft im Spannungsfeld zwischen Recht und Utilitarismus (1694–1715)* (Zurich: Schulthess, 1993)
Bazzoli, Maurizio, *Il piccolo Stato nell'età moderna: Studi su un concetto della politica internazionale tra XVI e XVIII secolo* (Milan: Jaca Book, 1990)
Bazzoli, Maurizio, *Stagioni e teorie della società internazionale* (Milan: LED, 2005)
Beguelin, Édouard, "En souvenir de Vattel (1714–1767)," in *Extrait du Recueil des travaux offerts par la Faculté de droit de l'Université de Neuchâtel à la Société suisse des Juristes à l'occasion de sa réunion à Neuchâtel 15–17 septembre 1929* (Neuchâtel: Université de Neuchâtel, 1929)
Blair, Ann, "Authorial Strategies in Bodin," in *The Reception of Bodin*, ed. Howell A. Lloyd (Leiden-Boston: Brill, 2008), 137–156
Bois, Jean-Pierre, *De la paix des rois à l'ordre des empereurs 1714–1815* (Paris: Editions du Seuil, 2003)
Boucher, David, *Political Theories of International Relations: From Thucydides to the Present* (Oxford: Oxford University Press, 1998)
Bozzola, Annibale, "Venezia e Savoia al congresso di Utrecht (1712–1713)," *Bollettino storico-bibliografico subalpino* XXXV 3–4 (1933), 30–39
Bozzola, Annibale, *Giudizi e previsioni della diplomazia medicea sulla Casa di Savoia durante la guerra di successione di Spagna* (Turin: G. Bonis e Rossi, 1914)
Buonomo, Vincenzo, "Reciprocità, libertà religiosa e protezione dei diritti umani in ambito internazionale," in *Libertà religiosa e reciprocità*, ed. Jose Antonio Arana Mesa (Milan: Giuffrè, 2009), 119–148
Calabria, Antonio, and Marino, John A., eds., *Good government in Spanish Naples* (Berlin-New York: Peter Lang, 1990)
Chambrier, Frédéric-Alexandre de, *Histoire de Neuchâtel et Valangin jusqu'à l'avènement de la Maison de Prusse* (Neuchâtel: Attinger, 1840)
Christov, Theodore, "Vattel's Rousseau: Jus Gentium and the Natural Liberty of States," in *Western Political Science Association 2011 Annual Meeting Paper*. Available at: https://ssrn.com/abstract=1766921, 167–169

Crespo Solana, Ana, "Cooperation or Neutrality? How War Affects Business Strategies: The Case of Cadiz, Spain (1700–1720)," in *Neutres et neutralité dans l'espace atlantique durant le long XVIIIe siècle (1700–1820)*, ed. Éric Schnakenbourg (Bécherel: Les Perséides, 2015), 31–58

Fiocchi Malaspina, Elisabetta, *L'eterno ritorno del Droit des gens di Emer de Vattel (secc. XVIII–XIX). L'impatto sulla cultura giuridica in prospettiva globale* (Frankfurt: Max Planck Institute for Legal History, 2017)

Gabba, Emilio, and Schiavone, Aldo (eds.), *Polis e piccolo Stato tra riflessione antica e pensiero moderno (atti della giornate di studio 21–22 febbraio 1997)* (Florence: New Press, 1997).

Good, Christoph, *Emer de Vattel (1714–1767). Naturrechtliche Ansätze einer Menschenrechtsidee und des humanitäre Völkerrecht im Zeitalter der Aufklärung* (Zurich: Dike Verlag AG, 2011).

Grandpierre, Louis, *Histoire du Canton de Neuchâtel sous le Roi de Prusse 1707–1848* (Leipzig: Grandpierre, 1805)

Hont, Istvan, *Jealousy of Trade: International Competition and the Nation-State in Historical Perspective* (Cambridge, MA: Harvard University Press, 2005)

Hurrell, Andrew, "Vattel: Pluralism and its Limits," in *Classical Theories of International Relations*, ed. Ian Clark and Iver Neumann (Houndmills-New York: Palgrave, 2001), 233–255

Ingerbritsen, Christine, ed., *Small states in International relations* (Reykjavik: University of Iceland Press, 2006)

Johnston, Douglas M., *Historical Foundations of World Order: The Tower and the Arena* (Leiden-Boston: Martinus Nijhoff, 2008)

Jouannet, Emmanuelle, "Les dualismes du Droit des gens," in *Vattel's International Law from a XXIst Century Perspective*, eds. Vincent Chetail and Peter Haggenmacher (Leiden: Brill, 2011), 133–150

Lapradelle, Albert Geouffre de, "Introduction" to Emer de Vattel, *Le Droit des Gens ou principes de la loi naturelle appliqués à la conduite et aux affaires des nations et des souverains* (Washington: Carnegie Institution, 1916), iii–lix

Lebeau, Christine, "Negotiating a Trade Treaty in the Imperial Context: The Habsburg Monarchy in the Eighteenth Century," in *The Politics of Commercial Treaties in the Eighteenth Century. Balance of Power, Balance of Trade*, eds. Antonella Alimento and Koen Stapelbroek (Basingstoke: Palgrave Macmillan, 2017), 349–369

Musi, Aurelio, "La nazione napoletana prima della nazione italiana," in *Nazioni d'Italia. Identità politiche e appartenenze regionali tra Settecento e Ottocento*, eds. Angela De Benedictis, Irene Fosi and Luca Mannori (Rome: Viella, 2012), 75–89

Nakhimovsky, Isaac, "Vattel's Theory of the International Order: Commerce and Balance of Power in the Law of Nations," *History of European Ideas* 33 (2007), 157–173

Neff, Stephen C., "Peace and prosperity: commercial aspects of peacemaking," *Peace treaties and International Law* in *European History. From the Late Middle Ages to World War One,* ed. Randall Lesaffer (Cambridge: Cambridge University Press, 2004), 365–381

Onnekink, David, "The treaty of Utrecht 1713," in *Peace was made here: the treaties of Utrecht, Rastatt and Baden 1713–1714,* eds. Renger de Bruin and Maarten Brinkman (Petersberg: Im Hof, 2013), 60–69

Onuf, Nicholas Grenwood. "Civitas Maxima: Wolff, Vattel and the fate of Republicanism," *American Journal of International Law* 88 (1994), 287–296

Pocock, John G. A., *Barbarism and Religion,* vol. 1, *The Enlightenments of Edward Gibbon, 1737–1764* (Cambridge: Cambridge University Press, 1999), vol. 2, *Narratives of Civil Government* (Cambridge: Cambridge University Press, 1999)

Remec, Peter Paul, *The Position of the Individual in International Law according to Grotius and Vattel* (The Hague: Martinus Nijhoff, 1960)

Roulet, Louis-Edouard, "Friedrich der Große und Neuenburg," in *Friedrich der Große in seiner Zeit,* ed. Oswald Hauser (Cologne-Vienna: Böhlau, 1987), 181–191

Sabbatini, Renzo, and Volpini, Paola, eds., *Sulla diplomazia in età moderna. Politica, economia, religione* (Milan: FrancoAngeli, 2011)

Schmidt-Voges, Inken, "Making Peace in Early Modern Europe," in *Peace was made here: the treaties of Utrecht, Rastatt and Baden 1713–1714,* eds. Renger de Bruin and Maarten Brinkman (Petersberg: Im Hof, 2013), 49–59

Schnakenborug, Eric, *Entre la guerre et la paix. Neutralité et relations internationales XVIIe-XVIIIe siècles* (Rennes: Presses Universitaires de Rennes, 2013),

Schnettger, Matthias, "*Kleinstaaten in der Frühen Neuzeit. Konturen eines Forschungsfeldes,*" Historische Zeitschrift 236 (2008), 605–640

Stapelbroek, Koen, "'The long peace': commercial treaties and the principles of global trade at the Peace of Utrecht," in: *The 1713 Peace of Utrecht and its enduring effects,* ed. A.H.A. Soons (Leiden: Brill, 2019), 93–119

Stapelbroek, Koen, "L'organisation du commerce international dans l'ombre d'Utrecht: les perspectives hollandaises du XVIIIe siècle," in *La paix d'Utrecht (1713): Enjeux économiques, maritimes et commerciaux,* eds. Lucien Bély, Géraud Poumarède and Guillaume Hanotin (Paris: Pedone, 2018), 475–502

Stapelbroek, Koen, "Universal Society, Commerce and the Rights of Neutral Trade: Martin Hübner, Emer de Vattel and Ferdinando Galiani," in: *Universalism in International Law and Political Philosophy,* ed. Petter Korkman and Virpi Mäkinen, *Collegium. Studies across Disciplines in the Humanities and Social Sciences* 4 (2008) (Helsinki: Helsinki Collegium for Advanced Studies, 2008), 63–89

Stapelbroek, Koen, and Trampus, Antonio, "The Legacy of Vattel's Droit des gens: Contexts, Concepts, Reception, Translation and Diffusion," in: *The Legacy of Vattel's Droit des gens: Contexts, Concepts, Reception, Translation and*

Diffusion, eds. Koen Stapelbroek and Antonio Trampus (Cham: Palgrave Macmillan, 2019), 1–25

Toyoda, Tetsuya, "Vattel's doctrine of national sovereignty in the context of Saxony Poland and Neuchâtel," in *Theory and politics on the law of nations: Political Bias in International Law Discourse of Seven German Court Councilors in the Seventeenth and Eighteenth Centuries* (Leiden: Brill, 2011), 161–190

Tuck, Richard., *The Rights of War and Peace: Political Thought and the International Order From Grotius to Kant* (Oxford: Oxford University Press: 1999)

Whatmore, Richard, and Kapossy, Béla, "Emer de Vattel's Mélanges de littérature, de morale et de politique (1760)," *History of European Ideas* 34 (2008), 77–103

CHAPTER 3

The Good Government: The Constituting and Constituted Nation

Vattel's *Droit des gens* implied an imagined outline of a long-term process of how states evolved and by consequence interacted. To understand the reasons for the interest and success of Vattel's work it is therefore important to have a clear idea of the conceptions and interpretations of the concept of 'good government' that prevailed during the modern era.

WHAT IS A GOOD GOVERNMENT? THEORETICAL ANSWERS TO AN ANCIENT QUESTION

During the early modern age, Europe was home to a number of small states, some the enduring legacies of the Medieval and Renaissance eras, distributed in a fairly homogeneous way between monarchical and republican forms of government. This was the picture painted—with a focus on Italy but a glance across the continent—by Machiavelli in *The Prince* and the *Discourses on the First Decade of Titus Livius*, and which remained largely unchanged until, in the eighteenth century, it came to be idealised by means of the concept of the 'small state' seen as the model of virtue, both in international relations and in forms of government, and often associated with republicanism and the myth of good government.[1] The concept was also taken up in the late eighteenth century by Kant to resolve the confrontation between politics and morality. Indeed, in *Perpetual Peace* he explained that in international relations harmony between politics and morality was only possible through a federal union of small states,

© The Author(s) 2020
A. Trampus, *Emer de Vattel and the Politics of Good Government*,
https://doi.org/10.1007/978-3-030-48024-0_3

because only a federation that had the elimination of war as its main aim was compatible with their freedom. Accordingly, the disappearance of a small state or its absorption by a larger one is an example of poor politics and immorality.

As Norberto Bobbio said, "The nagging question that recurs throughout the history of political thought is: 'What is the better art of government, the one based on the rule of men or on the rule of law?'" Starting from antiquity the different responses to this question constitute one of the most interesting and even fascinating instruments by which to understand intellectual history and the history of political thought.[2] The discussion on the better art of government was a central question during the early modern age, but it nevertheless should not be confused with another traditional but different question on what is the best form of government.

The difference between the two questions and their respective answers lies in identifying the person capable of effectively protecting man and citizen in civil life. While affirming that the primacy of the law protects the citizen from the whim of the bad ruler, the primacy of man protects him from an indiscriminate broad-brush application of the general rule, as long as the ruler is just. The first solution shields the individual from singularity, that is, from the arbitrariness of decisions, the second from the generality of prescription that would ultimately level men, making them uniform by precluding the recognition of ability and merit. The primacy of law presupposes good law, and the primacy of man presupposes a good ruler. The two solutions have often been placed head to head as if they constituted an absolute choice, a radical alternative, either/or. But in practice, and in history, they end up being interchangeable.

The primacy of the law is also based on the assumption that rulers are for the most part bad, in so far as they tend to use power for their own ends. Conversely, the primacy of man is based on the assumption of a good ruler, in line with an ideal that recalls the classical myth of the just and wise lawmaker. In Western history, when this idea of good government was being translated into a practical rule, the preferred solution was often that of the superiority of the rule of law, due to the mistrust or generally negative judgement of those who by chance or by merit (or a combination of the two, as Machiavelli wrote) were in a position of managing the affairs of states.

Two criteria in particular distinguish good government from bad: first, whether the government respects established laws, be they natural or divine, or respects norms of custom or the positive laws decreed by its predecessors that have become customs of the country. And second, whether the government is deemed arbitrary, deciding matters on a case-by-case basis outside of all established rules.

All classic political thought of the West, from the Middle Ages onwards, has been characterised by the belief that a good ruler is one who governs by respecting laws that he cannot freely control, either because they transcend him (such as those handed down by God) or because they are part of the natural order of things, or again because they were established by the collectivity through the constitution of the state. This idea was rooted long ago in the English constitution, and, from the late seventeenth century spread from England to enter by degrees into the juridical cultures of the continental states. This gave rise to the idea of the 'state of law,' that is, of the state whose fundamental principle is the subordination of all power—from the highest level to the lowest—to the law, through a system of guarantees, balances and controls over every aspect of government. Since the birth of written constitutions in the eighteenth century this has been called 'constitutionalism.'

Good government, especially in the classical sense of the government by good rulers, has often been regarded as incompatible with the principles of freedom proper to modernity and the liberal tradition. The reasons for this stem from a different interpretation of the premises of ancient and medieval thought: in particular, the ideal of good life, the absolute centrality of the common good and the idea of the superiority of anything that is in common over what is individual were seen as placing limits on personal liberties. A consequence of this approach can be seen in the eighteenth century in the so-called science of good government, the political and economic theory widespread in the eighteenth-century German and Austrian territories and best represented by Joseph von Sonnenfels. Good government in this interpretation, Sonnenfels explained, consists of the principles and rules by which the internal security of the state is established and maintained. These are based primarily on the obedience of the people to government: compliance that can be spontaneous or result from the impossibility of resistance. Good government in Sonnenfels's interpretation consisted of the principles by which the internal security and preservation of the state was established and maintained. According to Sonnenfels it consisted principally in the observation of the law and in carrying out all the activities of state within the limits of the law. This was a notably different conception from the classical republican concept of good government that was often regarded as incompatible with the principles of modern state. Moreover, good government, construed as the security of the state, thus consists principally of observing the law and of the ability to carry out all its activities within legal limits.[3] This particular structure can be recognised in the eighteenth-century German and Austrian territories and in a large part of the Italian peninsula still governed directly or indirectly by the Habsburg monarchy: from Lombardy to the Austrian Littoral and from Tuscany to the Kingdom of Naples. The Austrian idea of good government had thus become part of Italian political culture.

However, in the culture of the liberal democracies good government has come to mean the exercise of power by a ruler who keeps within established laws and uses his power to pursue the common good. Hence the link with Aristotelean thought, which became increasingly relevant during the nineteenth century and according to which it is better to be governed by the best laws than by the best men, not least because laws do not have the passions that inevitably affect human nature. As for the economy, the science of economics ("a branch of the science of a statesman or legislator," according to Smith) had come to argue that it is the market which ensures that a good government is not merely one with an abstract relation to justice, because the market is the sphere in which the practice of good government is applied and confirmed. This economic postulation gave rise to the development in the twentieth century of a liberal theory of good governance.[4] The central notion in that theory is that the common good placed limits on personal liberties.[5] Competition and discussion, in other words the market and the public sphere, are the pillars of this idea of good governance, which became a veritable constitutional project, always under construction and continually evolving.

Transcending the either/or logic of these questions, the concept of good government posited the primacy of the rule of law as the precondition for the 'good ruler,' who governed by respecting laws that he could not freely control outside the constitution of the state.[6]

THE MYTH OF GOOD GOVERNMENT AND ITS IMPACT ON MODERN POLITICAL CULTURE

The concept of 'good government,' which became central to Vattel's reflection on constitutionalism, was therefore known in the history of Western political culture from the classical to the modern age, but before the *Droit des gens* the term was used to describe a political practice, to denote a way of exercising sovereignty, and was not yet in itself a conceptual category. Today, however, we know that good government is also a concept, a category of politics that is by nature changeable and adaptable to different contexts as a result of values that can be very different from each other and mutable over time. In the *ancien régime* all of this was not yet clear, because the concept was still undergoing a slow formation and evolution. Even though we do not have systematic research on the history and success of this concept and the lexical expressions that were associated with it in the seventeenth and eighteenth centuries, it appears that the modern age, and in particular the period running from the so-called crisis of European conscience to the birth of the Enlightenment, is the most fertile field for recognising the changes of its meanings. The term 'good government' was used only rarely for much of the seventeenth century, and when it did appear it was often in connection with religious and spiritual questions, within the government of churches and the faith, and almost never secularised. In Catholic culture 'good government' referred to the internal discipline of religious orders, of the community of believers, and of church institutions. It therefore mostly had to do with an activity of social and institutional control, which over time came to involve policing in the form of prevention and repression.[7] Even in English culture, the expression was linked to the religious life and the control of forms of dissent between the seventeenth and eighteenth centuries.[8]

Gradually, in civil and political discourse the expression 'good government' began to assume its secularised form, principally as a synonym of political 'perfection' and 'perfect government,' to explain some of the characteristics of republicanism, the antidotes to tyranny and the guarantees of peace.[9] A very interesting example of the evolution of the phrase comes to us through the work of the Venetian historian Paolo Paruta.[10] In France, meanwhile, even as early as the seventeenth century there was a gradual shift in the use of 'bon gouvernement' from the religious to the political sphere, and in the *Journal des Sçavans*, for example, both religious and lay usages recurred frequently.[11] At the beginning of the eighteenth century genuine fondness for the term increased dramatically throughout much of Europe, particularly when there sprang up a general tendency to transform knowledge into 'codifications' and into real 'sciences' by giving words and concepts strictly defined meanings in relation to different contexts.[12] Friedrich Karl von Moser's essay *Der Herr und der Diener, geschildert mit patriotischer Freiheit* (1759) was translated into French with the title *Idée d'un bon gouvernement* (1761). Between 1761 and 1765 Gaspard de Réal de Curban published the eight volumes of his *Science du gouvernement, ouvrage de morale, de droit, et de politique, qui contient les principes du commandement & de l'obéissance*, while in 1768 Alexandre Auguste de Campagne published the *Principes d'un bon gouvernement*. And thus from the mid-eighteenth century the expression 'good government' imposed itself as a particularly effective slogan with which to convey secularised political ideas, as a kind of container that could be filled with very different content pertinent to the ongoing debate about the transformation and crisis of the *ancien régime*. Use of the term increased, as far as one can tell, in parallel with the development of reflections on sovereignty, on democracy, on the relationship between morality and law, and on the problem of how to make political representation effective and of how to establish the best possible relations between the governed and the governors, up to the point of influencing the question of the democratic deficit in the contemporary age.[13]

Many people now argue that we owe the most mature elaboration of the concept of good government with these characteristics to American political thought, to the experience of freedom verified by the 1787 constitution and the work of Thomas Jefferson. But in reality, the concept was already central in Vattel's work and, if we consider the influence that this had on American constitutional history, it cannot be ruled out that Jefferson actually took it directly from Vattel. To understand the centrality and success of Vattel's work it is therefore important to bear in mind above all the fact that he was among the first to appropriate this expression and modern political category and make it a key point of his philosophical and political thought, precisely at the moment when it became a fashionable phrase in the languages and cultures of the European continent.

There is, however, a second reason for Vattel's success that is connected to the concept of good government, and this is the fact that the new semantic container was adapted to an older type of content, to a tradition of political virtue and anti-despotism proudly claimed by the small states of the modern age in the dialectic between freedom and tyranny.[14] To see why the European system and the small states of the *ancien régime* were particularly fertile ground for the discussion of these concepts one need only look at the famous cycle of frescoes by Ambrogio Lorenzetti in Siena. Although the title under which the frescoes are generally known, *Allegory of Good Government*, is a result of nineteenth-century historiography,[15] this immense work in fact provided a complex synopsis of the concept of 'good government,' illustrating the set of virtues that prevented the city from falling into discord, exasperation and conflict. In the images, Peace needs Justice, which does not hesitate to rely, if necessary, on stringent Security. And if in the cultural politics of Siena the idea of good government was associated with the values of peace and justice and the civic virtue of the small state,[16] eighteenth-century readers of Vattel preserved the same connections and used the *Droit des gens* and its conception of good government to make the domestic tradition of political thought fit for modernity.

The frescoes of the so-called good government thus presented in the Europe of the *ancien régime* a republican theory of civil government that revolved around the concept of the common good (*bonum commune, commune utilitates*, according to pre-humanist writers), of concord and of *aequitas* as the foundation of the *civitas*.[17] And this referred not only to the community of citizens since in the eyes of contemporaries the government was also able to interpret and guard those values. This category could also adapt well, following the classical tradition, to a small state or a republic, but, as Vattel himself explained, nothing prevented it being adopted by a good prince guided by the principles and rules of natural law.

We can grasp how powerful and how richly evocative this message was from the fact that soon—between the eighteenth and nineteenth centuries—the conceptional definition and theoretical description of good government would no longer be enough. It required some form of exemplification. Hence in the nineteenth century the idea of good government not only needed to be conceptualised but also to be illustrated visually, and it was in this context that the Siena frescoes on the allegory of war and peace would be given the name *Good and Bad Government and their effects*. This was the name that would be given to Lorenzetti's fresco cycle in the context of the Risorgimento and with the unification of Italy.[18]

Vattel on Good Government: The Internal Constitution of State and the Constituting Power as Political Will

If one central political question was whether the rule of men or the rule of law was to be preferred, it had to be distinguished from the question of which was the best form of government. These questions were at stake in the eighteenth-century Italian and Mediterranean contexts in which Vattel's *Droit des gens* was used as an instrument.[19]

It is striking from this perspective to see that Vattel referred to the same idea of politics as Sonnenfels, a normative conception of the constitutional state and of good government. The latter concept, to which Vattel devoted many pages of the *Droit des gens*, is central to grasping the commonalities and divergences between the classical tradition, natural-law thinking and the transition to political modernity. It chimed particularly well with the reformist drive that existed in the Mediterranean region and the Italian small states in particular in the later eighteenth century. Specifically, the concept of good government offered the possibility to contemplate the forming and reforming of new laws without the necessity to change the form of government.

Discussing good government in Book I, *Of Nations considered in themselves*, Vattel stated that:

> The fundamental regulation that determines the manner in which the public authority is to be executed, is what forms the *constitution of the state*. In this is seen the form in which the nation acts in quality of a body-politic,—how and by whom the people are to be governed,—and what are the rights and duties of the governors.[20]

According to Vattel, 'good government' was a corollary of this constitution and of "the perfection of a state." Subsequently, "its aptitude to attain the ends of society, must then depend on its constitution." Chapter VI was dedicated to the *Principal Objects of a good Government; and first to provide for the Necessities of the Nation* and showed what was meant by a state being perfected constitutionally and in its policies and laws. Here, Vattel stated that the "first object of a good Government is to provide the necessities of the Nation." It was the task of "good government" to produce a "happy plenty of all the necessaries of life" by managing the labour market, preventing emigration and fostering a healthy climate for industriousness in the fields of agriculture and domestic and foreign trade, and in monetary policy.[21]

The *Second Object of a good Government*, following Vattel, that is, *to procure the true Happiness of the Nation*, was connected with how 'good government' went beyond the provision of physical needs. Vattel argued that the state had to "instruct the people to seek felicity [...] in their own perfection." Therefore the state had a duty "to teach them the means of obtaining it," in much the same way as "the education of youth" deserved "the attention of the government" given that "literature and the polite arts" served to "enlighten the mind, and soften the manners." Moreover, it was "necessary to inspire the people with the love of virtue, and the abhorrence of vice." These feelings were the foundation of a perfect patriotism that kept the citizens of the nation united by being seen to be mutually advantageous to them all and thereby to the state: "The grand secret of giving to the virtues of individuals a turn so advantageous to the state, is to inspire the citizens with an ardent love for their country."[22]

Then, in Chapter XIV Vattel rehearsed the same scheme, as he did several times in the *Droit des gens*, but replaced the discussion of good government in the internal constitution of the state with the concept of good government in the interstate system: "The third Object of a good Government [is] to fortify itself against external attacks. One of the ends of political society is to defend itself with its combined strength against all external insult or violence. [...] To increase the number of the citizens as far as it is possible or convenient [...] The wealth of a nation constitutes a considerable part of its power, especially in modern times."[23] It was in this way that good government not only provided a solvent for the historical-political contradictions that eighteenth-century Italian political reformist thought ran into, but also a guideline for the reconsideration of the position of the Italian small states in the competitive international arena of the time.

This process was accomplished, Vattel claimed, when, in order to become relevant on the international stage, a state or nation had to become a *constitutional state*, with all that this entailed. A constitution exists when there are laws made specifically with the public good in mind, such as those that relate to the body itself and the essence of society or the fundamental laws relating to the form of government and the organisation of powers: "those, in a word, which together form the constitution of the state, are the *fundamental laws*."[24]

Vattel also postulated—and was perhaps the first in Europe to do so—the distinction between *constituting power as political will* and *constituted power*.[25] As a consequence, "a nation has an indisputable right to form, maintain, and perfect its constitution," while, as regards legislative powers, "the nation may intrust the exercise of it to the prince, or to an assembly; or to that assembly and the prince jointly."[26]

This type of constitutional state was therefore the real subject destined to act in international politics to safeguard the natural rights of society and of individuals from within the system that presupposed the attainment of formal equality of all states, large and small. Moderation in the conduct of war (*ius in bello*) ensues from this, as does the notable importance attributed to the purpose of neutrality. The state must first promote the common good of its citizens, basing its authority on the consent of the governed and recognising the right of people to choose their own laws without external interference. Needless to say, these were arguments that could be adapted to serve the reform policies of eighteenth-century Europe, in particular those aimed at advancing general wellbeing and public happiness.[27] The common good of a nation, according to Vattel, was directly linked to the role of trade and political economy, which were not meant to support politics of power but to look after the primary needs of the population, and agriculture rather than manufacturing. He believed that domestic monopolies damaged the rights of citizens but accepted the use of monopolies at the international level—as a means of protecting a nation's trade—as long as they were restricted to the need of sustaining the life of a population.[28]

As we might note, this insistence on the relationship between state and nation and on the internal constitution of the states corresponded to the idea that the main aim of the *Droit des gens* was to provide a logic for the rationalisation of interstate conflict.[29] While this historiographical judgement of Vattel's *Droit des gens* is typical of the work's reception ever since its publication, it fails to do justice to Vattel's views on the resolution of interstate conflict. The solution to these conflicts, in fact, could only come after all the problems arising from the internal political organisation of the states had been resolved and clarified. In fact, Vattel directly engaged with the key political challenges of his time by very narrowly circumscribing war as a situation in which a state rightly pursued its laws by strength,[30] and he tried to align this notion of war as a sanctionative juridical instrument with a framework for the reform of European politics.[31]

The cornerstone of Vattel's political thought was the idea of legal autonomy, which connected his notions of individual self-preservation and perfection to the concept of the state. On this basis, each society of men that had united itself politically was, according to natural law, free and sovereign, irrespective of what degree of power or political and economic autonomy it possessed. Even when placed under the protection of another state, or in a situation of military inequality, a state remained free and sovereign. Vattel famously reasoned that states, as societies of men, stood in the same relation to one another with regard to their rights and obligations as individuals within any state.[32]

In this way Vattel forged a shift away from the focus on dynastic, economic and military interests of Europe's dominant territorial monarchies and their struggle for hegemony and placed the wellbeing of society in the legal foreground.[33] Ultimately, it was this shift, which followed from the basic principles of Vattel's political thought, that would turn him into a leading light for politicians of small states who sought to break free of the idea that the only hope of survival consisted either in maintaining perfect passive neutrality and political moderation aimed at self-preservation, or in seeking refuge within a hegemonic state.

Parallel to the idea that the main aim of the *Droit des gens* was to provide a logic for the rationalisation of interstate conflict,[34] the work offered an outlook on interstate trade rivalry. This is how its discussion of commercial neutrality is best understood as being more than an arbitrary compromise between the neutral right to trade and the belligerent right to intercept trade out of necessity. In line with his idea of perfection and regulated luxury, Vattel advocated a law of neutrality that separated trade in higher-end (mistakenly presumed to be directly 'perfecting') goods from the more essential and fundamental needs-based trade in subsistence goods.[35] The first kind of trade remained subject to all the natural rights that states had to interfere with each other's commerce, while the second was to be considered an inviolable realm of natural, politically neutral exchange between individuals.

Vattel also discussed the legal restrictions that ought to exist with regard to (private) investment in foreign state debts by citizens of neutral states. He emphasised the idea that such investments should not in any way be politically interested (even if such investments led to spending on warfare); in other words, interest payments had to be in line with market conditions.[36] In the course of dealing with the politics of commercial relations between citizens of different states, and by opening up questions about obligations and rights emanating from treaty arrangements, Vattel discussed political neutrality and trade in relation to each other more than any writer had done before, in order to respond to the problem of neutral trade in the War of the Austrian Succession.

Just as the Kingdom of Naples exemplified a medium-sized state much closer to the status of a small one than that of a great power, so Austrian Lombardy, where Pietro Verri and the intellectuals of the *Il Caffè* magazine worked, was an example of a region with a medium-sized territory and limited sovereignty constantly torn between the demands of old political freedoms and the rule of the House of Habsburg. More generally, because of the chronic weakness of the Holy Roman Empire and the Habsburgs' efforts to transform their fragmented and uneven territorial units into a modern state, it was the Austrian monarchy, with all its Italian possessions, that proved best suited to the discussion of Vattel's ideas. It can therefore be understood why the *Droit des gens* was so widely read and used as a textbook both in the universities of Austrian territories—beginning in Vienna[37]—and by writers in Habsburg lands, as in the case of Sonnenfels from the 1760s.[38] The publication and diffusion of Vattel's book coincided with a change in European international politics after the repositioning of Spain in the continent and in the Mediterranean area, and with a profound transformation of the strategic alignments in Habsburg foreign policy (the "renversement des alliances") planned by Kaunitz in the context of the Seven Years' War. This was a response to the failures of the War of the Austrian Succession and renewed conflict with Prussia, which led Kaunitz to abandon the old notion of power politics and of a European balance of power and to see an alliance with France as the starting point for a true system of peace ("ein wahres Friedens Systema") based on the preservation of common interests.[39] A direct influence of Vattel's thinking is discernible in this stance, especially in the assertion that the European states, despite their opposing interests, were closely dependent on one another and that anyone trying to impair this international order should be considered a public enemy.[40] It was also at the time of the Seven Years' War that Kaunitz changed his understanding of and the meaning of the term 'convenience' (*droit de convenance*), which up to that point had been linked to the outcome of the War of the Spanish Succession[41] and the concept of interest associated with Pufendorf.[42]

Vattel, having been a diplomatic envoy to Dresden, was known to the Habsburg court and was cited in the correspondence between Maria Theresa and Princess Maria Christina of Saxony,[43] Maria Theresa's fourth daughter and the wife from 1766 of Albert of Saxony-Teschen. It is therefore easy to see why his work and the *Droit des gens* in particular circulated so widely in the Habsburg monarchy and served to explain the key role that this country, with its Italian offshoots, would come to play in the new international economic system of the 1760s.

NOTES

1. Maurizio Bazzoli, *Il piccolo Stato nell'età moderna. Studi su un concetto della politica internazionale tra XVI e XVIII secolo* (Milan: Jaca Book, 1990), 11–51; Christine Ingerbritsen, ed., *Small states in International relations* (Reykjavik: University of Iceland Press, 2006); Domokos Kosary, *Les "petits Etats" faceaux changements culturels, politiques et économiques de 1750 à 1914* (Lausanne: HU Jost, 1985); Dieter Langewiesche, ed., *Kleinstaaten in Europa: Symposium am Liechtenstein-Institut zum Jubiläum 200 Jahre Souveränität Fürstentum Liechtenstein 1806–2006* (Vaduz: Verlag der Liechtensteinischen Akademischen Gesellschaft, 2007).
2. Norberto Bobbio, *The Future of Democracy: A Defence of the Rules of the Game* (Minneapolis: University of Minnesota Press, 1987), 138.
3. The reference is to Joseph von Sonnenfels's concept of "gute Policey" (good order), translated in eighteenth-century Italy as "the science of good government" (*Scienza del buon governo scritta dal Signor di Sonnenfels e recata dal Tedesco in italiano* (Milan: Giuseppe Galeazzi, 1784). On Sonnenfels's political thought and his idea of good government and good order, see Keith Tribe, *Governing Economy: The Reformation of German Economic Discourse 1750–1840* (Cambridge: Cambridge University Press, 1988), 55–90; László Kontler, "Polizey and Patriotism: Joseph von Sonnenfels and the Legitimacy of Enlightened Monarchy in the gaze of eighteenth-century State Sciences," in: *Monarchism and Absolutism in Early Modern Europe*, ed. Cesare Cuttica and Glenn Burgess (London and New York: Routledge, 2012), 82–83.
4. One of the major international exponents of which was the Italian economist Luigi Einaudi, who later became president of the Italian Republic.
5. Bobbio, *The Future of Democracy*, 138; Robert Heineman, *Authority and Liberal Tradition from Hobbes to Rorty* (New Brunswick-London: Transaction Publishers, 1994), 72.
6. The concept of 'good government' (from the French word 'governement') should not be confused with that of 'good governance' (from the English word 'governance'), which is a term that was not yet widespread in the eighteenth century. In this I follow Jean-Pierre Gaudin and Yu Keping, according to whom "Literally, there seems no great difference between

'governance' and 'government.' Yet semantically, they are vastly different. To many scholars, a prerequisite for correct understanding of governance is to distinguish it from government. As Jean-Pierre Gaudin said, 'Governance has to be distinguished from the traditional concept of government by the State from the very beginning.' As a political administration process like government by the State, governance also requires authority and power and ultimately aims to maintain a normal social order. Despite their similarities, there are two fundamental differences between them" (Jean-Pierre Gaudin, "Modern governance, yesterday and today: some clarifications to be gained from French government policies," *International Social Science Journal* 50 (155) (1998): 47–56; Yu Keping, "Governance and Good Governance: A Framework for Politica Analysis," *Fudan Journal of the Humanities and Social Science* 11 (1) (March 2018): 1–8.

7. On this point see Filippo Sabetti, ed., *Alla ricerca del buon governo in Italia* (Manduria: Lacaita, 2004); Stefano Tabacchi, *Il Buon governo: le finanze locali nello Stato della Chiesa (secoli XVI–XVIII)* (Rome: Viella, 2007); Chiara Lucrezio Monticelli, *La polizia del papa: istituzioni di controllo sociale a Roma nella prima metà dell'Ottocento* (Soveria Mannelli: Rubbettino, 2012).

8. See John Newman, *The Character and Blessings of a Good Government: A Sermon Preach'd at Salters-Hall November 5th 1716* (London: Richard Ford, 1716); Robert Pattison, *The Great Dissent: John Henry Newman and the Liberal Heresy* (Oxford: Oxford University Press, 1991).

9. Simonetta Adorni Braccesi and Mario Ascheri, eds., *Politica e cultura nelle Repubbliche italiane dal Medioevo all'età moderna. Firenze, Genova, Lucca, Siena, Venezia* (Rome: Istituto storico italiano per l'età moderna e contemporanea, 2001); Gino Benzoni, "La città del Buon Governo: Venezia," in *Il Buono e il Cattivo Governo. Rappresentazioni nelle Arti dal Medioevo al Novecento*, ed., Giuseppe Pavanello (Venice: Marsilio, 2004).

10. Marco Giani, "Paolo Paruta: il lessico della politica" (doctoral thesis, Ca' Foscari University, 2011), https://www.academia.edu/1604301/Paolo_Paruta_Il_lessico_della_politica; see also Quentin Skinner, *The Foundation of Modern Political Thought*, vol. 1, *The Renaissance* (Cambridge: Cambridge Uiversity Press, 1998).

11. See "Extrait de diverses pièces envoyées pour étreines par Mr. Bernier à Madame de la Sabilère," *Le Journal des Sçavans*, 7 June 1688, 17–26; *Supplement du Journal des Sçavans*, January 1707, 357.

12. Rosario Patalano and Sophus A. Reinert, "Introduction," to *Antonio Serra and the Economics of Good Government* (New York: Palgrave Macmillan, 2016), 3.

13. Pierre Rosanvallon, *Le bon gouvernement* (Paris: Editions du Seuil, 2016); Tommaso Edoardo Frosini, "Is good government a myth?" in Sabetti, ed., *Alla ricerca del buon governo*, 5–10.
14. Umberto Mazzone, *'El buon governo:' un progetto di riforma generale nella Firenze savonaroliana* (Florence: Olschki, 1978).
15. Rosa Maria Dessì, "L'invention du 'Bon Gouvernement.' Pour une histoire des anachronismes dans les fresques d'Ambrogio Lorenzetti (XIVe–XXe siècle)," *Bibliothèque de l'Ecole de Chartes* 165 (2007): 129–180.
16. Quentin Skinner, "Ambrogio Lorenzetti's buon governo frescoes: Two old questions, two new answers," *Journal of the Warburg and Courtauld Institutes* 62 (1998): 1–8; Quentin Skinner, *Visions of Politics*, vol. II, *Renaissance Virtues* (Cambridge: Cambridge University Press, 2002), 39–92.
17. Quentin Skinner, "Ambrogio Lorenzetti: The Artist as Political Philosopher," *Proceedings of the British Academy* 72 (1986), 1–6; Quentin Skinner, "Il Buon governo di Ambrogio Lorenzetti e la teoria dell'autogovero repubblicano," in Braccesi and Ascheri, *Politica e cultura nelle repubbliche italiane*, 21–42.
18. Dessì, "L'invention du 'Bon Gouvernement'"; Rosa Maria Dessì, "Il bene comune nella comunicazione verbale e visiva. Indagini sugli affreschi del 'Buon Governo,'" in *Il Bene comune: forme di governo e gerarchie sociali nel basso medioevo* (XLVIII Convegno storico internazionale, Todi, 12–15 ottobre, 2011) (Spoleto: Centro italiano di studi sul l'alto medioevo: 2012), 89–130.
19. Antonio Trampus, *Storia del costituzionalismo italiano nell'età dei Lumi* (Rome-Bari: Laterza, 2009). See also Pietro Costa, "The Rule of Law: A Historical Introduction," in *The Rule of Law: History, Theory and Criticism*, ed., Pietro Costa and Danilo Zoro (Dordrecht: Springer, 2007), 73–150.
20. Vattel, *Droit des gens*, book I, ch. III, § 4, 26.
21. Vattel, *Droit des gens*, book I, ch. VI, § 7–276.
22. Vattel, *Droit des gens*, book I, ch. XI, § 110–124.
23. Vattel, *Droit des gens*, book I, ch. XIV, § 177–186.
24. Vattel, Droit des gens, book I, ch. III, § 29.
25. Vattel, *Law of nations*, book I, ch. III, § 31 and 34.
26. Emer De Vattel, *Il diritto della natura e delle genti ovvero principii della legge naturale, opera recata nell'italiano da Lodovico Antonio Loschi*, vol. 1 (Lyon [but Venice: Vitto], 1781), 40.
27. John Christian Laursen, Hans Blom and Luisa Simonutti, eds., *Monarchism in the Age of Enlightenment. Liberty, Patriotism and Common Good* (Toronto: University of Toronto Press, 2007).

28. Isaac Nakhimovski, "Vattel's theory of the international order: Commerce and the balance of power in the *Law of Nations,*" *History of European Ideas* 33 (2007): 157–173, 168–170.

29. Isaac Nakhimovsky, "Carl Schmitt's Vattel and the *Law of Nations* between Enlightenment and Revolution," *Grotiana* 31 (2010).

30. For instance, Vattel, *Droit des gens,* book II, ch. III, § 7. See Simone Zurbuchen, "Vattel's law of nations and just war theory," *History of European Ideas* 35 (2009): 408–417.

31. See Nakhimovsky, "Vattel's theory of the international order," 157–173; Koen Stapelbroek, "Universal Society, Commerce and the Rights of Neutral Trade: Martin Hübner, Emer de Vattel and Ferdinando Galiani," *COLLeGIUM: Studies Across Disciplines in the Humanities and Social Sciences* 3 (2008): 63–89.

32. Bazzoli, *Il piccolo Stato nell'età moderna,* 15; Ingerbritsen, *Small states in International relations;* Domokos Kosary, *Les "petits Etats" faceaux changements culturels, politiques et économiques de 1750 à 1914* (Lausanne: HU Jost, 1985); Langewiesche, ed., *Kleinstaaten in Europa.*

33. Vattel, *Droit des gens,* Preliminaries, § 18.

34. Nakhimovsky, "Carl Schmitt's Vattel."

35. For the argument that Vattel's views on the interstate system derived from his ideas on luxury see Nakhimovsky, "Vattel's theory of the international order."

36. Vattel, *Droit des gens,* book III, ch. VII, § 110.

37. Gigliola Di Renzo Villata, *Formare il giurista: esperienze nell'area lombarda tra Sette e Ottocento* (Milan: Giuffrè, 2004).

38. Tribe, *Governing economy,* 79.

39. Lothar Schilling, *Kaunitz und das Renversement des alliances. Studien zur aussenpolitischen Konzeption Wenzel Anton Kaunitz* (Berlin: Duncker und Humblot, 1994), 335.

40. Vattel, *Droit des gens,* book III, ch. XV, § 222 and book IV, ch. I, § 5.

41. Schilling, *Kaunitz,* 336.

42. Friedrich Kratochwil, "On the notion of "interest" in international relations," *International organizations* 36 (1982): 1–30.

43. Woldermar Lippert, *Kaiserin Maria Theresia und Kurfuerstin Maria Antonia von Sachsen: Briefwechsel, 1747–1772* (Leipzig: Teubner, 1903).

BIBLIOGRAPHY

Most of the quotations of Vattel's *Droit des gens* come from the widely available English-language edition of 2008, which includes an introduction by Béla Kapossy and Richard Whatmore (Indianapolis: Liberty Fund) and maintains the text and English title *The Law of Nations: Or, Principles of the Law of Nature*

Applied to the Conduct and Affairs of Nations and Sovereigns of the 1797 London standard edition. Contemporary translations of primary sources are listed under the names of their authors. Full manuscript sources are referenced only in the notes for reasons of space.

PRIMARY SOURCES

JOURNALS

Le Journal des Sçavans, 7 June 1688 (Paris: Académie des Inscriptions & belles-lettres, 1688)
Supplement du Journal des Sçavans, January 1707 (Paris: Delespine, 1707)

PRINTED BOOKS

Newman, John, *The Character and Blessings of a Good Government: A Sermon Preach'd at Salters-Hall November 5th 1716* (London: Richard Ford, 1716)
Sonnenfels, Joseph von, *Scienza del buon governo scritta dal Signor di Sonnenfels e recata dal Tedesco in italiano* (Milan: Giuseppe Galeazzi, 1784)
Vattel, Emer de, *Il diritto della natura e delle genti ovvero principii della legge naturale applicati alla condotta e agli affari delle nazioni e de' sovrani scritta nell'idioma francese dal Sig. di Vattel e recata nell'italiano da Lodovico Antonio Loschi* (Lyon [Venice]: n.p. [Vitto], 1781)
Vattel, Emer de, *Law of Nations: Or, Principles of the Law of Nature, Applied to The Conduct and Affairs of Nations and Sovereigns*, eds. Béla Kapossy and Richard Whatmore (Indianapolis: Liberty Fund 2008)

SECONDARY SOURCES

Adorni Braccesi, Simonetta, and Ascheri, Mario (eds.), *Politica e cultura nelle Repubbliche italiane dal Medioevo all'età moderna. Firenze, Genova, Lucca, Siena, Venezia* (Rome: Istituto storico italiano per l'età moderna e contemporanea, 2001)
Bazzoli, Maurizio, *Il piccolo Stato nell'età moderna: Studi su un concetto della politica internazionale tra XVI e XVIII secolo* (Milan: Jaca Book, 1990)
Benzoni, Gino, "La città del Buon Governo: Venezia," in *Il Buono e il Cattivo Governo. Rappresentazioni nelle Arti dal Medioevo al Novecento*, ed. Giuseppe Pavanello (Venice: Marsilio, 2004), 93–108
Bobbio, Norberto, *The Future of Democracy: A Defence of the Rules of the Game* (Minneapolis: University of Minnesota Press, 1987)

Costa, Pietro, "The Rule of Law: A Historical Introduction," in: *The Rule of Law: History, Theory and Criticism*, eds. Pietro Costa and Danilo Zoro (Dordrecht: Springer, 2007), 73–150

Dessì, Rosa Maria, "Il bene comune nella comunicazione verbale e visiva. Indagini sugli affreschi del "Buon Governo"," in *Il Bene comune: forme di governo e gerarchie sociali nel basso medioevo* (XLVIII Convegno storico internazionale, Todi, 12–15 ottobre, 2011) (Spoleto: Centro italiano di studi sul l'alto medioevo: 2012), 89–130

Dessì, Rosa Maria, "L'invention du 'Bon Gouvernement.' Pour une histoire des anachronismes dans les fresques d'Ambrogio Lorenzetti (XIVe–XXe siècle)," *Bibliothèque de l'Ecole de Chartes* 165 (2007), 129–180

Di Renzo Villata, Gigliola, *Formare il giurista: esperienze nell'area lombarda tra Sette e Ottocento* (Milan: Giuffrè, 2004)

Frosini, Tommaso Edoardo, "Is good government a myth?," in *Alla ricerca del buon governo in Italia*, ed. Filippo Sabetti (Manduria: Lacaita, 2004), 5–10

Gaudin, Jean-Pierre, "Modern governance, yesterday and today: some clarifications to be gained from French government policies," *International Social Science Journal* 50 (155) (1998), 47–56

Giani, Marco, "Paolo Paruta: il lessico della politica" (doctoral thesis, Ca' Foscari University, 2011)

Heineman, Robert, *Authority and Liberal Tradition from Hobbes to Rorty* (New Brunswick-London: Transaction Publishers, 1994)

Ingerbritsen, Christine, ed., *Small states in International relations* (Reykjavik: University of Iceland Press, 2006)

Keping, Yu, "Governance and Good Governance: A Framework for Politica Analysis," *Fudan Journal of the Humanities and Social Science* 11 (1) (March 2018), 1–8

Kontler, László, "Polizey and Patriotism: Joseph von Sonnenfels and the Legitimacy of Enlightened Monarchy in the gaze of eighteenth-century State Sciences," in *Monarchism and Absolutism in Early Modern Europe*, eds. Cesare Cuttica and Glenn Burgess (London and New York: Routledge, 2012), 75–90

Kosary, Domokos, *Les "petits Etats" faceaux changements culturels, politiques et économiques de 1750 à 1914* (Lausanne: HU Jost, 1985)

Kratochwil, Friedrich, "On the notion of "interest" in international relations," in *International organizations* 36 (1982), 1–30

Langewiesche, Dieter, ed., *Kleinstaaten in Europa: Symposium am Liechtenstein-Institut zum Jubiläum 200 Jahre Souveränität Fürstentum Liechtenstein 1806–2006* (Vaduz: Verlag der Liechtensteinischen Akademischen Gesellschaft, 2007)

Laursen, John Christian Laursen, Blom, Hans and Simonutti, Luisa, eds., *Monarchism in the Age of Enlightenment. Liberty, Patriotism and Common Good*, (Toronto: University of Toronto Press, 2007)

Lippert, Woldermar, *Kaiserin Maria Theresia und Kurfuerstin Maria Antonia von Sachsen: Briefwechsel, 1747–1772* (Leipzig: Teubner, 1903)

Mazzone, Umberto, *'El buon governo:' un progetto di riforma generale nella Firenze savonaroliana* (Florence: Olschki, 1978)

Monticelli, Chiara Lucrezio, *La polizia del papa: istituzioni di controllo sociale a Roma nella prima metà dell'Ottocento* (Soveria Mannelli: Rubbettino, 2012)

Nakhimovsky, Isaac, "Carl Schmitt's Vattel and the *Law of Nations* between Enlightenment and Revolution," *Grotiana* 31 (2010), 141–164

Nakhimovsky, Isaac, "Vattel's Theory of the International Order: Commerce and Balance of Power in the Law of Nations," *History of European Ideas* 33 (2007), 157–173

Patalano, Rosario, and Reinert, Sophus A., *Introduction*, to *Antonio Serra and the Economics of Good Government* (London: Palgrave Macmillan, 2016)

Pattison, Robert, *The Great Dissent: John Henry Newman and the Liberal Heresy* (Oxford: Oxford University Press, 1991)

Rosanvallon, Pierre, *Le bon gouvernement* (Paris: Editions du Seuil, 2016)

Sabetti, Filippo, ed., *Alla ricerca del buon governo in Italia* (Manduria: Lacaita, 2004)

Schilling, Lothar, *Kaunitz und das Renversement des alliances. Studien zur aussenpolitischen Konzeption Wenzel Anton Kaunitz* (Berlin: Duncker und Humblot, 1994)

Skinner, Quentin, "Ambrogio Lorenzetti: The Artist as Political Philosopher," *Proceedings of the British Academy* 72 (1986), 1–6

Skinner, Quentin, "Ambrogio Lorenzetti's buon governo frescoes: Two old questions, two new answers," *Journal of the Warburg and Courtauld Institutes* 62 (1998), 1–8

Skinner, Quentin, "Il Buon governo di Ambrogio Lorenzetti e la teoria dell'autogoverno repubblicano," in *Politica e cultura nelle Repubbliche italiane dal Medioevo all'età moderna. Firenze, Genova, Lucca, Siena, Venezia*, ed. Simonetta Adorni Braccesi and Mario Ascheri (Rome: Istituto storico italiano per l'età moderna e contemporanea, 2001), 21–42

Skinner, Quentin, *The Foundation of Modern Political Thought*, vol. 1, *The Renaissance* (Cambridge: Cambridge University Press, 1978)

Skinner, Quentin, *Visions of Politics*, vol. 2, *Renaissance Virtues* (Cambridge: Cambridge University Press, 2002)

Stapelbroek, Koen, "Universal Society, Commerce and the Rights of Neutral Trade: Martin Hübner, Emer de Vattel and Ferdinando Galiani," in: *Universalism in International Law and Political Philosophy*, ed. Petter Korkman and Virpi Mäkinen, *Collegium. Studies across Disciplines in the Humanities and Social Sciences* 4 (2008) (Helsinki: Helsinki Collegium for Advanced Studies, 2008), 63–89

Tabacchi, Stefano, *Il Buon governo: le finanze locali nello Stato della Chiesa (secoli XVI–XVIII)* (Rome: Viella, 2007)

Trampus, Antonio, *Storia del costituzionalismo italiano nell'età dei Lumi* (Rome-Bari: Laterza, 2009)

Zurbuchen, Simone, "Vattel's law of nations and just war theory," *History of European Ideas* 35 (2009), 408–417

CHAPTER 4

The First Reception: Sicily, Corsica and the Mediterranean Islands

The use made of the *Droit des gens* in the specific context of the Mediterranean islands is worthy of particular interest. Indeed these territories were in some respects very similar to Vattel's home country, Switzerland, in so far as many of them, especially Sicily and Corsica, which were under the rule of Spain and France respectively, were strongly oriented towards policies of economic and constitutional reform and the search for a new role on the international scene.

THE MEDITERRANEAN AND VATTEL'S *DROIT DES GENS*: THE CASE OF SICILY

The Mediterranean in the eighteenth century was not only the region of small states, but also one that was trying to deal with huge international transformations by creating new forms of sovereignty and engaging in competitive trade. As such, it represented the perfect terrain for the dissemination of Vattel's theories. Yet a second reason for focusing on the Mediterranean region consists in the fact that Sicily, Corsica and more generally the Italian peninsula were places which, by being slow to discard the legacy of the feudal world due to the absence of economic reforms, eventually became particularly lively political laboratories for European

---

Archival Abbreviations
BP   Bibliothèque Patrimoniale de Bastia, Bastia, France

© The Author(s) 2020   59
A. Trampus, *Emer de Vattel and the Politics of Good Government*,
https://doi.org/10.1007/978-3-030-48024-0_4

reformist policies and for the construction of a political language destined to influence the constitutional and democratic development of the whole world: one thinks of Pascal Paoli, Cesare Beccaria, Pietro Verri and Gaetano Filangieri.

Sicily represents the intriguing first case for the diffusion of Vattel's work. The island, along with the Spanish possessions in southern Italy, had gone from being a direct dominion of the Bourbons of Spain to an autonomous kingdom under the infant don Carlos, who was crowned Charles III in 1735 in the cathedral of Palermo. The existence of a Sicilian crown formally autonomous from that of Naples implied a recognition of local juridical and political traditions and institutions, beginning with the island's ancient parliament. When the new king received the crown of Sicily with the approval of the parliament and from the hands of the Archbishop of Palermo, the sovereign's independence with respect to the Pope and the diplomatic policy of the Holy See was symbolically underlined. Twenty years later, in 1759, just when Vattel had finished writing the *Droit des gens*, Ferdinand VI of Spain died and Charles III was called to the Spanish throne, providing an ideal opportunity for testing on the scene of the international balances what had been theorised in the aftermath of the Utrecht peace. To comply with political and dynastic tradition, Charles III had to unite the Sicilian and Neapolitan crowns with the Spanish crown, but international treaties expressly forbade this union. The impasse was broken when he had his eight-year-old third-born son, Ferdinand, made king with the name Ferdinand I, establishing a regency council controlled by the minister Bernardo Tanucci to support him.

This is one of the earliest occasions in which the *Droit des gens* was used in a debate on practical questions concerning the political and constitutional organisation of a state. After the death of Charles III of Bourbon, when Sicily faced the danger of reunification to the Spanish crown and thus needed to establish the legitimacy of its independence and sovereignty, a young jurist from Girgenti (now Agrigento) got to work using Vattel in support of his arguments. The man in question was Vincenzo Gaglio (1735–1777), a law graduate from Palermo who practised law in Catania and studied history, literature and archaeology.[1] Gaglio was a member of the Accademia del Buon Gusto in Palermo, which had been founded in 1718 in honour of Ludovico Antonio Muratori[2] and occupied itself with literature and the sciences, as well as the study of law, to the extent that it even considered establishing its own private law school.[3] It was also associated with the jurisdictional question, that is, with protecting the sovereign from the Pope's attempts to exert influence on him.

The publication in 1759 of the *Saggio sopra il diritto della natura e delle genti e della politica. Dell'avvocato Vincenzo Gaglio girgentino, accademico del Buon-Gusto*[4] was not therefore the result of an unconsidered decision by its author. Rather, Gaglio's aim was not only to intervene in the debate on the foundations of natural law and the relationship between law and morality, but also to reflect on the problem of sovereignty, on the relationship between the prince and the autonomy of the island, and on the relationship between the law of nature and the royal right to administer the internal civil and political life of Sicily. As stated in his own account, his twin goal was to educate young people in the study of public law—a science still neglected in Italy—and to ensure the good governance of cities, republics and societies.

For this reason his work was divided, following a structure not unlike that of the *Droit des gens*, in two main parts. The first was dedicated to the foundations of natural law, in which Gaglio developed an in-depth critique of the ideas of Grotius, Hobbes and Spinoza (in Pope's *Essai sur l'homme* translated by Étienne de Silhouette), of Pufendorf through the newly published Italian translation by Giambattista Almici, of Locke's *Essay philosophique*, and of Burlamaqui's *Principes du droit naturel* and Pufendorf's *Devoirs de l'homme* (in Barbeyrac's translation). Using arguments typical of the contractualism of natural law and the idea of man's natural sociability and inherent understanding of right and wrong, Gaglio launched first of all into a critique of Machiavelli and his contention that laws were made only to serve the princes and not for the public good.[5] Since instead the function of laws was public benefit, it followed, according to Gaglio, that they must have had an inspiring principle superior to the will of the prince, identifiable in the inalterability of natural law. The consequences therefore consisted in the intrinsic goodness of this law and in the natural existence of man's duties with regard to its observance, and to respect of God and society, as Barbeyrac and Pufendorf, as well as Malebranche and Cumberland, had affirmed.[6]

However, Gaglio did not bring Vattel into play until the second half of his work, starting in Chap. 5, where he used him to criticise the conception of the law of nations proposed by Grotius. Gaglio held that it was in fact wrong to conceive of a positive law of nations distinct from natural law. The mere consent of nations, in other words a purely voluntary law that some believed would have brought about the so-called law of nations, did not in fact produce real obligatory effects. For this reason, it was instead necessary to argue for the existence of a superior law to regulate trade between the nations, and this could only be the law of nature, the same law that was subsequently called the law of nations when applied to peoples and nations. This type of natural law of nations appeared to be compulsory because it recognised nature and God as its primary source and because it obliged nations trading with each other to follow the same practices, and to observe the same duties to which private citizens were subjected.[7]

It is easy to discern what practical requirement gave birth to Gaglio's reasoning. With regard to a purely voluntary law of nations, such as that established by treaties that could easily be violated, the only way to affirm the inviolable nature of Sicilian sovereignty was, he believed, to appeal to the eternal and unchangeable natural law. He went so far as to give some concrete examples of the consequences of this reasoning, which were important for understanding the geopolitical situation of Sicily. One concerned maritime sovereignty, the other the function of diplomacy.

On the first point Gaglio refuted Bynkershoek' *Questions of Public Law* (1737),[8] in the part of his work where he argued that, in keeping with natural law, property coincided with possession so that when one loses possession of something by not keeping it the thing becomes the property of the initial owner. Applying this principle to the dominion of the sea, the problem concerned the waters around Sicily and the strategies to avoid the danger of the island losing sovereignty over them when occupied and temporarily dominated by a foreign power.

As for the function of diplomacy and the right of each nation to practise it autonomously, Gaglio rejected Grotius's idea that diplomacy was born of the law of nations, in that it was the product of simple convention, and instead maintained that it derived directly from natural law because it was this that induced people to seek the peace necessary for the advance of commerce. Consequently, since the ambition of the Sicilian nation was the development of trade and the maintenance of peace, then under natural law the island's parliament could autonomously maintain relations with other nations and other states.

The final part of the work tackled the question of which was the best form of government.[9] While recognising that the most natural way to organise power was that of the monarchy, Gaglio claimed that natural law fulfilled the function of inspiring and guiding the authority of the prince and of subjecting this to limits and conditions by listing his duties towards his subjects. This preference for monarchy was justified by the fact that a democratic government could, in the normal way of things, raise to the highest offices citizens who enjoyed the highest level of support but who often turned out to be the least capable of governing. The intrinsic weakness of mixed governments, on the other hand, was the influence of external powers that laid bare their lack of stability.[10]

The importance of Vattel's work (again through its first edition of 1758) to the Sicilian debate was documented by Gaglio in an essay dealing with the subject of a medical controversy and published twenty years later. Entitled *Lettera al sig. Pepi sull'estrazione del feto vivente e morboso* [i.e. *ammalato*] *nei parti pericolosi e difficili*, it had been inspired by a debate launched by a doctor from Agrigento, Giovanni Carbonaro (or Carbonajo), about the opportuneness of saving the life of a mother during a difficult parturition when the unborn child's life was at risk, and therefore about whether to prefer aborting the unborn when it was deemed to be ill or capable of endangering the life of the mother. Carbonara had been answered by another doctor, Antonino Pepi, who used the natural law theories of Christian Wolff to justify the killing of the foetus in such a case and described the unborn as an unjust aggressor who deserved that sentence. The question led to a lively debate among the island's intellectuals, who contributed various texts, all collected in a volume called *Opuscoli di autori siciliani* (vol. 19, Palermo 1778).[11]

Gaglio's contribution of almost 100 pages made extensive use of Vattel. He declared that it was necessary to move beyond Wolff's philosophy: "I recommend the reading of the great treatise on the law of nations written with great clarity by Mr. Vattel, who according to the authors of the *Novelle letterarie* in Bern has with his insights overtaken Wolff himself."[12] Making use of the culture of natural law and also Vattel, in this case the second book of the *Droit des gens* especially, Gaglio compared the fight for survival by a mother and unborn child during a critical birth to that of two warring nations. He pointed out that the position of the unborn was in fact and by nature a weaker and dependant one, just as in the law of nations in the case of a competition between a powerful nation and a weaker one that depends on it. This being the case, it follows, as demonstrated by nature itself, that the foetus, because it is weaker, can never be considered an assailant of the mother and a threat to her, in just the same way as a weaker nation cannot be considered the aggressor of a stronger one. It was the weaker party that had the right to defend itself, and the position of the weakest should in every case be preserved.[13]

Similarly, the recourse to illicit weapons or other disproportionate means must always be considered taboo in spite of the natural right to self-defence: "Ask Vattel," Gaglio wrote. "This barbarity will thunder in your ears. It is contrary to the principles of the law of nature and the rights of humanity, because it is not permissible to add to the evils of war and make war itself more cruel and frightening."[14] Gaglio therefore wondered whether anything forbidden in war could be allowed during a difficult and dangerous childbirth. Was not the right to self-defence equal for the weak nations and the unborn child? His reasoning concluded by demonstrating the illegitimacy of the preventive termination of an unborn child, even if it were ill and could put the life of the mother at risk. In his long essay Gaglio most ably introduced into the clinical question not only basic notions of eighteenth-century natural law, but also the corollary of Vattel's rationale that judged ending the life of an unborn child to be a manifestation of violence. For natural law a foetus could not be considered "an unjust aggressor," and, ergo, did not deserve to die. Moreover, for evangelical morality it must also be loved as a neighbour and a weakling, and destined therefore to be delivered. "If it is not lawful," Gaglio ended, "according to the wise opinion of Vattel, to add to the evils of war, it is also not permissible to add to the evils of humanity."[15]

Insular Freedom Against the Continent: Vattel in Corsica

Another of the small states, or one such aspirant, in which intensive use was made of Vattel was Corsica during the years of its revolution against Genoa and the beginning of French domination. The source that enables us to gain greater understanding of this context is an annotated copy of the *Droit des gens* preserved in the *Bibliothèque patrimoniale* of Bastia. This is a copy of the Dutch edition of 1758, bound in high-quality parchment and which, although not bearing the owner's name, appears to have come from the library of a scholar or official.[16] The fact that the copy was used for study or work is suggested by the inclusion of numerous annotations of obvious eighteenth-century origin which, for reasons outlined below, may be assigned to the period immediately after the ceding of the island to France in 1769. The annotations are not composed of handwritten notes or comments but of simple marks running through the entire volume, being mainly either vertical lines or crosses. Evidently, the book's owner used these signs to mark the level of importance of passages as he read them, the crosses being employed when a more immediate and specific referral was needed.

The annotation of readings of Vattel's work and the possible uses made of it in the Corsican political debate is not surprising, particularly if one thinks of the intense intellectual exchanges then being made between Corsica and the Italian peninsula as well as between Corsica and France and England in the years of the revolution. The exchange of ideas with the Naples of Bernardo Tanucci is especially well-documented in the writings of Antonio Genovesi,[17] while, with particular regard to the *Droit des gens*, the close connections between the constitutional experiences of Corsica and of Tuscany that coincided with the first attempts to translate Vattel's work into Italian are already known.[18]

The annotator employs two different levels of reading: the first concerns the overall structure of the work, whose essentially political dimension is outlined by the general picture given by the index of chapters at the start of the first volume. More than Vattel's introduction, the Corsican reader appears to have thought that this was the best way of showing the work's relevance to the ongoing public debate. The annotation and the significance attributed to different chapters and their contents is expressed through hurried pen strokes, essential reminders, marks and highly visual pointers. The second, more in-depth, level of interpretation is that relating to the content of individual sections. Here we find many other, smaller and more detailed signs in the margins of particular parts of the text. The purpose of these annotations seems to be different, being aimed more at a specific, practical use relating to contingent events. This appears to be confirmed by the fact that the process of the annotations and their reference points in the two levels of annotation—that of contents and that of text—is not symmetrical and the points of interest do not always coincide: on more than one occasion, for example, a reference mark placed in the margin of the section and chapter titles listed in the general index does not correspond to a similar sign in the respective contents, and vice versa. It is therefore worth reconstructing the logic of the unknown reader to consider in particular how he interpreted the overall structure of the *Droit des gens*.

The reader's attention seems to have been focused in equal measure on the parts of the work that deal with the constitutional dimension of the state and more specifically with the law of nations and of international society. The methods of annotation demonstrate a clear interest in the theory of sovereignty outlined by means of the general layout of the work, as well as in the formation process of sovereign authority with regard to the exercise of power by the prince and the rights of majesty.[19] Despite the simplicity of the annotative solution used to denote the importance of the various themes, the reader manages to establish ties between different parts of the text by using longitudinal lines that highlight and link various points of argument. For example, an unbroken line links the exercise of sovereignty to the existence of a pact based on compromise between the prince and his subjects,[20] and to the independence of the territory from any foreign influence.[21] Further on, more underscoring indicates the parts that deal with how to procure the true happiness of a nation, in other words the value of educating young people and of punishing crimes against the homeland.[22] Particular attention is given to the paragraphs on the rights of citizens in the event of a nation being subject to a foreign power,[23] to those on citizenship, to cases where a citizen is forced to leave his homeland,[24] and to the possibility of a sovereign violating the rights of emigrants.[25] Finally, at the end of the first book, the sections referring to the complex relationship between the control of public property as an exercise of sovereignty and the protection of private property are also marked.[26]

In the indexes of the second book by Vattel, which is dedicated to relations between nations, the reader underlines the chapters relating to the sovereign's obligation to protect citizens,[27] to the freedom of families to live independently in a country without necessarily depending on the organisation of the political society,[28] to the fact that the state has no jurisdiction over foreign citizens and their possessions,[29] and finally to the fact that there are rights which men cannot be deprived of, in particular that of continuing to live in a territory that has become part of a foreign country.[30]

The Interpretation of International Treaties

There are also many marks alongside the titles of sections devoted to the interpretation of international treaties. Here attention appears to shift towards more familiar aspects of Vattel's contribution to the discussion of international society, but once again these relate to themes closely connected to the key to reading propounded in the first part. This appears clear, for example, from the emphasis—once again indicated with marks—placed on the question of the invalidity of treaties dangerous to the state,[31] on the incompatibility of treaties and pre-existent natural duties,[32] and on the fact that anything that one could not or chose not to make clear in treaties remains to the detriment of the contractors.[33] Once again attention was placed on the interpretation of obscure clauses of treaties, on the extensibility of this interpretation, and on the fact that interpretations aimed at a common usefulness, equality, human society and the preservation of the status quo were preferable, while interpretations that ran counter to them were unacceptable.[34] Finally, all the sections relating to the illegality of reprisals against foreigners residing within a state were duly noted.[35]

The volume does not contain any clear indication that makes it possible to identify the annotator or to make a secure dating of this reading. Nevertheless, the analysis carried out so far at least allows us to narrow, with sufficient reliability, the timeframe to a particular moment, since all the elements highlighted here indicate that the annotator made use of the *Droit des gens* in the period following the conclusion of the Treaty of Versailles between Genoa and France of 15 May 1768, and probably not long after the incursion of May 1769 that consolidated the French presence on the island. It was therefore a historical phase in which Vattel's work—which on the one hand satisfied all the requirements of authority and, on the other, referred (for reasons already mentioned) to a context of ideologically neutral production—avoided imbalances in favour of one of the international powers deployed on the field. The anonymous reader could therefore have been an inhabitant of the island seeking to understand whether the Corsican nation would be able to fit into the new institutional and state system, and with what liberties and guarantees, as well as wanting to know how widely the Treaty of Versailles might be applied and interpreted.

If we enter further into the text of the *Droit des gens* to analyse other traces of this reading, we may observe how the underscoring is particularly dense in the sections to do with the characteristics, rights and obligations of the nation, and also in the one which affirms the principle that all nations are free and independent in nature, just as men are.[36] The annotator commented on the fact that there was a customary right of nations that competed with positive law,[37] that the nation was born out of a natural compulsion and could exist and survive independently of the state and its transformations.[38] Consequently, only the nation itself had the right to change its constitution and choose its form of government,[39] and a sovereign had a duty to understand the nation or nations within his state.[40] Considerable underscoring appears to have been reserved for the claims that a people must not—since it was contrary to the rights of a nation—accept foreign laws (here the reference is to the very passage that in the 1774 edition of the *Droit des gens* would be completed by an explicit reference to the relations between Pope Gregory VII and Corsica).[41] Similarly, the unknown Corsican reader considered important the passages in which Vattel stressed the importance of long-term possession and usucaption as a source of natural law by which a nation could acquire ownership of a territory.[42]

The annotations more or less end at this point, although the margins of the third book still contain some markings in the form of a third sign, a double cross used to draw attention to sections on the justification of war,[43] on illegitimate war,[44] and primarily on the treatment of the property of those deemed enemies and the question of compensation for those who suffered losses due to war.[45] If, as it seems, these signs are another element useful for the dating of the reading, then the notes could be placed in a period shortly after 1769, in a limited span of time that included 1772 in which—once the dream of the Corsican kingdom as a small state in the European system had faded—the island still remained an occupied territory and the problem of war reparations was widely discussed.[46]

NOTES

1. Gaetano Di Carlo, *Vincenzo Gaglio, Un intellettuale nella Girgenti del Settecento tra giurisprudenza e archeologia:* (doctoral thesis, University of Pisa, 2015–2016), https://core.ac.uk/download/pdf/79621731.pdf; Antonio Palermo, *Vincenzo Gaglio e il Rinnovamento siciliano* (Agrigento: Siculgrafica, 2017). On the cultural context, see Nicola Cusumano, *Libri e cultura in Sicilia nel Settecento* (Palermo: New Digital Press, 2016), especially 18–20.

2. Vincenzo Di Giovanni, "Le origini delle accademie dei Riaccesi e del Buon Gusto (1568, 1622, 1718)" in *L'accademia nazionale di scienze lettere e arti di Palermo, 1718–1984: note storiche pubblicate in onore dei partecipanti alla 59. sessione della Union Academique Internationale riunita a Palermo dal 2 all'8 giugno 1985* (Palermo: Accademia di Scienze Lettere e Arti di Palermo, 1985), 19–54.

3. Domenico Scina, *Prospetto della storia letteraria di Sicilia nel secolo decimottavo* (Palermo: Lo Bianco, 1859), vol. 1, 18–22; Giuseppe Giarrizzo, "Appunti per la storia culturale della Sicilia settecentesca," in *Rivista Storica Italiana* 79 (1967), 584–587.

4. The short book was published in Palermo, by the printer Francesco Valenza in 1759. For this research I have used the copy preserved in Palermo in the Sicilian Regional Library, under Antiqua III 1004/III 78.
5. Gaglio, *Saggio sopra il diritto della natura e delle genti*, 1–15.
6. Gaglio, *Saggio sopra il diritto della natura e delle genti*, 17–25.
7. Gaglio, *Saggio sopra il diritto della natura e delle genti*, 45.
8. Gaglio, *Saggio sopra il diritto della natura e delle genti*, 51, the reference is to Cornelis van Bynkershoek, *De dominio maris dissertatio* (1702), in *Opera minora* (Leiden: van de Kerckhem, 1744), ch. 1.
9. Gaglio, *Saggio sopra il diritto della natura e delle genti*, 99–115.
10. Gaglio, *Saggio sopra il diritto della natura e delle genti*, 103–198, with reference to Bodin.
11. In the collection Gaglio had already published an essay on the "Problema storico, critico politico se la Sicilia fu più felice sotto il governo della Repubblica romana o sotto isuoi imperatori" (1776).
12. Vincenzo Gaglio, "Lettera al sig. Pepi sull'estrazione del feto vivente e morboso ne' parti pericolosi e difficili," in *Opuscoli di autori siciliani*, vol. 19 (Palermo: Rupetti, 1778), 25–115, in particular 45. There is a reference to this controversy in Santi Correnti, "Avvisaglie femministe nel Settecento," in *La Sicilia nel Settecento*, ed. Francesco Renda (Messina: Centro di Studi Umanistici, 1986), 183.
13. Gaglio, "Lettera al sig. Pepi sull'estrazione del feto vivente," 80–82.
14. Gaglio, "Lettera al sig. Pepi sull'estrazione del feto vivente," 92–93, with a specific reference to the *Droit des gens*, book III, ch. VIII, § 156.
15. Gaglio, "Lettera al sig. Pepi sull'estrazione del feto vivente," 94.
16. BP, P-75-6.2. This was a single book binding together both volumes of the work.
17. The picture given by Franco Venturi, *Settecento riformatore*, vol. V/1, *L'Italia dei Lumi (1764–1790)*, (Turin: Einaudi, 1987), 4–201, remains unequalled.
18. For an analytical comparison between the events of Corsican constitution-alism and the troubled publication of Peter Leopold's constitutional plan, in light of the central role of the political vocabulary introduced by the *Droit des gens*, see my essay "Tra Corsica e Toscana: Emer de Vattel e i percorsi del costituzionalismo settecentesco," *Etudes Corses* 78 (2014): 61–80.
19. Vattel, *Droit des gens*, book I, ch. I, § IV, 45–46.
20. Vattel, *Droit des gens*, book I, ch. IV, § 52.
21. Vattel, *Droit des gens*, book I, ch. V, § 67.
22. Vattel, *Droit des gens*, book I, ch. XI, § 111 and 123.
23. Vattel, *Droit des gens*, book I, ch. XVI, § 195.
24. Vattel, *Droit des gens*, book I, ch. XIX, § 212, 220, 223.

25. Vattel, *Droit des gens*, book I, ch. XIX, § 226.
26. Vattel, *Droit des gens*, book I, ch. XX, § 244, 245, 252, 254; ch. XXI.
27. Vattel, *Droit des gens*, book II, ch. VI, § 71.
28. Vattel, *Droit des gens*, book II, ch. VII, § 97.
29. Vattel, *Droit des gens*, book II, ch. VIII, § 108–109.
30. Vattel, *Droit des gens*, book II, ch. IX, § 116, 125.
31. Vattel, *Droit des gens*, book II, ch. XII, § 160.
32. Vattel, *Droit des gens*, book II, ch. XII, § 170.
33. Vattel, *Droit des gens*, book II, ch. XVII, § 264.
34. Vattel, *Droit des gens*, book II, ch. XVII, § 284, 299, 301, 302, 305.
35. Vattel, *Droit des gens*, book II, ch. XVIII, § 342, 345, 348.
36. Vattel, *Droit des gens*, Preliminaries, 5.
37. Vattel, *Droit des gens*, book I, ch. I, § 27.
38. Vattel, *Droit des gens*, book I, ch. II, § 14–16.
39. Vattel, *Droit des gens*, book I, ch. II, § 31–34.
40. Vattel, *Droit des gens*, book I, ch. IV, § 44, 65.
41. Vattel, *Droit des gens*, book I, ch. IV, § 147.
42. Vattel, *Droit des gens*, book I, ch. XI, and ch. XXII, § 266.
43. Vattel, *Droit des gens*, book III, ch. III, § 29, 33.
44. Vattel, *Droit des gens*, book III, ch. IV, § 66, 67; ch. XI, § 183.
45. Vattel, *Droit des gens*, book III, ch. V, § 73, 76; ch. IX, § 168; ch. XIII, § 196–200; ch. XV, § 232.
46. See the letters written by Alexandre-Louis-Gabriel de Roux to his father in Christine Roux, *Les 'Makis' de la resistance corse 1772–1778* (Paris: France Empire, 1984), 25–56.

BIBLIOGRAPHY

Most of the quotations of Vattel's *Droit des gens* come from the widely available English-language edition of 2008, which includes an introduction by Béla Kapossy and Richard Whatmore (Indianapolis: Liberty Fund) and maintains the text and English title *The Law of Nations: Or, Principles of the Law of Nature Applied to the Conduct and Affairs of Nations and Sovereigns* of the 1797 London standard edition.

Contemporary translations of primary sources are listed under the names of their authors. Full manuscript sources are referenced only in the notes for reasons of space.

ARCHIVAL ABBREVIATIONS

BP   Bibliothèque Patrimoniale de Bastia, Bastia, France

PRIMARY SOURCES

PRINTED BOOKS

Bynkershoek, Cornelis van, *De dominio maris dissertatio* [1702], in *Opera minora* (Leiden: van de Kerckhem, 1744)

Gaglio, Vincenzo, "Lettera al sig. Pepi sull'estrazione del feto vivente e morboso ne' parti pericolosi e difficili,"in *Opuscoli di autori siciliani,* vol. 19 (Palermo: Rupetti, 1778), 25–115

————, *Saggio sopra il diritto della natura e delle genti e della politica. Dell'avvocato Vincenzo Gaglio girgentino, accademico del Buon-Gusto* (Palermo: Valenza, 1759)

Vattel, Emer de, *Law of Nations: Or, Principles of the Law of Nature, Applied to The Conduct and Affairs of Nations and Sovereigns,* eds. Béla Kapossy and Richard Whatmore (Indianapolis: Liberty Fund 2008)

SECONDARY SOURCES

Correnti, Santi, "Avvisaglie femministe nel Settecento siciliano," in *La Sicilia nel Settecento,* ed. Francesco Renda (Messina: Centro di Studi Umanistici, 1986), 129–132

Cusumano, Nicola, *Libri e cultura in Sicilia nel Settecento* (Palermo: New Digital Press, 2016)

Di Carlo, Gaetano, *Vincenzo Gaglio, Un intellettuale nella Girgenti del Settecento tra giurisprudenza e archeologia* (doctoral thesis, University of Pisa, 2015–2016)

Di Giovanni, Vincenzo, "Le origini delle accademie dei Riaccesi e del Buon Gusto (1568, 1622, 1718)," in *L'accademia nazionale di scienze lettere e arti di Palermo, 1718–1984: note storiche pubblicate in onore dei partecipanti alla 59. sessione della Union Academique Internationale riunita a Palermo dal 2 all'8 giugno 1985* (Palermo: Accademia di Scienze Lettere e Arti di Palermo, 1985), 19–54

Giarrizzo, Giuseppe, "Appunti per la storia culturale della Sicilia settecentesca," in *Rivista Storica Italiana* 79 (1967), 573–627

Palermo, Antonio, *Vincenzo Gaglio e il Rinnovamento siciliano* (Agrigento: Siculgrafica, 2017)

Roux, Christine, *Les 'Makis' de la resistance corse 1772–1778* (Paris: France Empire, 1984)

Scina, Domenico. *Prospetto della storia letteraria di Sicilia nel secolo decimottavo,* vol. 1 (Palermo: Lo Bianco, 1859)

Trampus, Antonio, "Tra Corsica e Toscana: Emer de Vattel e i percorsi del costituzionalismo settecentesco," *Etudes Corses* 78 (2014), 61–80

Venturi, Franco, *Settecento riformatore,* vol. 5/1, *L'Italia dei Lumi (1764–1790),* (Turin: Einaudi, 1987)

CHAPTER 5

The Great Crisis of the Sixties and the Political Reforms Between Piedmont and Tuscany

In general, not much attention has been given to the fact that the *Droit des gens* enjoyed a new lease of life after the death of its author in 1767. In a century, like the eighteenth, in which the modern conception of copyright was still developing, writers' deaths gave added impetus to publishing piracy, reworkings, updates and even distortions of their work. In the case of the *Droit des gens*, the period following the death of Vattel coincided, in the West, with the most acute phase of the crisis of the *ancien régime*, with the most innovative experiences from an economic and political point of view, with the prelude to the so-called Atlantic Revolutions and, in consequence, with the tendency to want to trace and read in the philosophical, political and economic works of those who had already disappeared prophetic auguries of the new era that was opening.

It is therefore not surprising that the *Droit des gens* was at the centre of a publishing resurgence at this time, with new editions and interpretations

---

Archival Abbreviations

ASF   Archivio di Stato di Firenze, Florence, Italy
ASM   Archivio di Stato di Modena, Modena, Italy
AST   Archivio di Stato di Torino, Turin, Italy
AV   Archivio Verri, Fondazione Raffaele Mattioli, Milan, Italy
FQS   Fondazione Querini Stampalia, Venice, Italy
HHStA   Haus-, Hof- und Staatsarchiv Wien, Vienna, Austria
UA   Universiteitsbibliotheek Amsterdam, Amsterdam, The Netherlands

© The Author(s) 2020   73
A. Trampus, *Emer de Vattel and the Politics of Good Government*,
https://doi.org/10.1007/978-3-030-48024-0_5

appearing. And these were not mere reprints, but rather brand new editions that presented the text in a new graphic layout, in new interpretative formulas, and with an articulation of the contents such as to induce different readings of the past, often adding passages purported to be from unpublished material of the author, or which were apocryphal, or were written by others to serve as clarification and guidance for readers of the text. This was the case with the edition printed in Neuchâtel in 1773 that was said to be "augmented" compared to the 1758 one thanks to the enterprise of the owners of the *Société typographique*.[1] The 1777 edition, meanwhile, had an introductory *Abrégé de la vie de M. de Vattel* which declared that the *Droit des gens* "will always be seen by connoisseurs as a work of the first order, intended to enlighten nations on their most essential interests."[2]

These new editions are interesting because they display certain changes of perspective in the use of Vattel's work. As we shall see in this chapter, through a wider cultural and political way of looking at the challenges of Europe's long-standing trade republics, the old theme of the relation between Vattel's political theory and the fate of small nations in eighteenth-century Europe was re-addressed. Ultimately, in contrast to the idea of Vattel as the last great theorist of natural law, the question is to what extent actual usage of the *Droit des gens* helped to generate international law.

Small and Medium-Sized States in the International Politics of the 1770s and 1780s

Up to this time, Italian interest in the part of the *Droit des gens* relating to international relations was either slight or totally absent. A notable exception was an article published by Alessandro Verri in *Il Caffè* (no. 32, 1766) which contained a reflection on the system of international politics and the legitimacy of wars of conquest. Alessandro Verri (who had pronounced legal interests and had written about Grotius) eulogised Vattel at length, describing him as one of the first writers to add depth to international political science by explaining "the true interests of nations." He thus used Vattel to criticise Machiavelli's political realism and to support the need for an international methodology which used wellbeing and trade, and not the force of arms, as the criteria for estimating the greatness of the states. Restating Vattel almost word for word, he declared that Europe had already become one great nation in which there must no longer be a place for international intrigues, dynastic interests and the ambition of conquest.[3] Another fairly isolated echo of Vattel's work (though not a specific mention) can be found in the *Lezioni di commercio* (1768) by Antonio Genovesi, who addressed the problem of the function of the small state, explaining that small political bodies manage to survive only precariously and invited consideration of what role could be played by a medium-sized state ("medio Stato"). Evidently he had in mind the Kingdom of Naples, which, though territorially large, was hardly a power in the international arena. Through Vattel, therefore, Genovesi adapted the theory of equality among nations to medium-sized states characterised by a reasonable ratio of population to "right territorial extension," as indicated by contemporary populationist theories.[4]

The group of intellectuals associated with the *Il Caffè* magazine in Milan were among the first people in Italy to recognise the importance of this reflection. Pietro Verri, who owned the first edition of the *Droit des gens*, used it from 1761 to 1762 in the first drafts of his *Saggio della grandezza e decadenza del commercio di Milano* to present an idea of trade based on free competition and the reduction of privileges and monopolies. Since the driving force of individual initiative is self-love, the art of the legislator should consist of directing the private interests of citizens towards the prosperity of the state. Vattel thus became an authoritative source for Verri's contention that trade pursued with rigour and force never develops, while trade based on kindness, justice and good laws does.[5] It was in that assessment that the notion of so-called absolute and relative goodness of laws emerged, an idea that would have great success in eighteenth-century Italy. The concept, in fact, dated back to the treatises of the late sixteenth century, and was revisited in the eighteenth century by Montesquieu, as well as being echoed in the *Droit des gens*. Later, in the manuscript of the second part of the *Stato attuale del commercio di Milano* (1762), Verri borrowed almost verbatim a large passage from Vattel that underlined the economic interdependence of European states at a time when Europe was seen as one nation made up of different families rather than a summation of different peoples.[6] He used the natural law arguments found mostly in a section of the *Droit des gens* dealing with voluntary natural law, purposefully focusing on the first volume of Vattel's work, which discussed the principles and organisation of the state and drawing in particular on the passages that recognised the individual's freedom to renounce his ties with political society, and therefore with the nation. This freedom, he declared, forces the government to commit itself to strengthening ties within society through incentives and through the recognition of individual prerogatives, especially economic ones, so as to allow a perceptible sense of homeland, justice and protection under the law.[7] Once more making a literal translation of Vattel, Verri also made clear that trade was a common good of the nation that was irreconcilable with every kind of monopoly and from which all citizens should rightfully benefit.[8] These assertions apparently did not conflict either with what he wrote about the duty of the prince to ensure the preservation of society and about his right to raise taxes (in this case, on salt in Milan) to pay troops to guarantee stronger protection against external threats, or with his defence of the right of the prince to dispose of public goods despite being simply their administrator rather than their owner.[9]

As time passed Vattel remained a reference point for Verri's work, being used in both theoretical reflection and government debate. Between 1769 and 1770 he was cited as an authority, together with Grotius and Pufendorf, in a memorandum sent by Verri to the government of Lombardy to claim just compensation for those of its subjects whose money gifts it had arbitrarily expropriated.[10] In the same period, he used Vattel in the manuscript annotations that he wrote in the margins of the edition of the *Meditazioni sulla economia politica* that appeared in Venice with a comment by Gianrinaldo Carli (1771). He referred to the *Droit des gens* yet again with regard to free trade, in the face of opposing opinions as to how extensive the "regulatory" role of the prince should be in relation to society. Carli, who inclined towards absolutist reformism, conceived the legislation and governmental control of the economy as a useful disciplinary tool to maintain the subjugation of the people. In contrast, Verri recalled the "spirit" of Vattel, which he believed coincided with that of the "heart of the princes of Europe," to maintain "order, peace, regularity" and "to trade as freely as possible and to remove the least possible freedom from artisans and merchants."[11]

A similar positioning can be found in the *Dei delitti e delle pene* by Cesare Beccaria, and this is not surprising. The text was drafted at the same time as Verri's economic reflections and was a product of the arguments and discussions that the two authors had shared. Beccaria held that the basis of the powers of political authority was contractual and therefore that the need to defend communal wellbeing from individual demands and abuses justified the right to punish. He again made reference to the first book of the *Droit des gens* in order to clarify that the legitimacy of punishments and the foundations of the right to impose penalties were grounded in the principle of the conservation of society.[12] Indeed, modern commentators have rightly pointed out numerous implicit references and allusions to Vattel's work throughout Beccaria's essay.[13] Often the use of the *Droit des gens*, which was again limited to the first volume, became more implicit, but it is important to note that, like Verri, Beccaria and the *Il Caffè* group used Vattel to adhere to a theory of sovereignty that, because of its roots in natural law, appeared superficially to offer support to the enlightened despotism and absolutism of the Habsburgs, but in reality prepared the way for a profound renewal of society. It was via this route that, through the translations of *Dei delitti e delle pene*, the *Droit des gens* gained a new and wide readership in Spain, and many admirers of Beccaria clamoured for the abolition of the death penalty, endorsing the *umanissimo* Vattel, as for instance did Alberto De Simoni, a lawyer from Valtellina and author of *Del furto e sua pena* (1776).[14] The diffusion of Vattel's work in Lombardy then and in the following years has been well-documented: for instance, lawyers of the Fiscal College used it when, on 26 March 1776, they responded to Kaunitz's request for their view on the advisability of abolishing torture.[15]

The gradual predominance of commercial interests over dynastic interests in European policy was felt sharply by the small states that found themselves at an economic disadvantage. In 1768, prior to the Austrian policy of Trieste, the government of the Republic of Venice had already bemoaned the start of a war "of industry" and "of commerce" against its manufacturers, with serious consequences for its sea trade with Puglia.[16] The situation compelled the Republic to rethink and to attempt to reformulate the policy of "exact neutrality" that Europe had pursued up to that time. Andrea Tron, a major exponent of institutional life in Venice and a former ambassador to The Hague, Paris and Vienna, followed this line in his studies. His works of those years, most probably influenced by his reading of Vattel, showed how he endeavoured to comprehend the true nature of the European system and of the new concept of interests. He concluded that Venice must retain its old policy of neutrality after completely overhauling it. This was therefore transformed from passive or "perfect" neutrality into active neutrality, no longer linked to contingent situations but directed in an ethical way so that, in its neutral status, the state used the benefits of trade to strengthen its independence and prestige.[17] The archives of Venetian families contain many traces of the readings carried out between the 1750s and 1770s of texts on European international politics, on the consequences of the Seven Years' War and on the theoretical problem of the just war. An important example due to the fact that it concerns the Querini family, which had roles of great importance in the government of the Republic, is the Italian translation—without, however, any added comments—of the *Réflexions d'un Suisse sur les motifs de la guerre présente* (1757) by Jean-Henri Maubert de Gouvest, an author close to the circles of the *Observateur hollandois* and to Moreau and Vattel, who was financed by the French government.[18]

In this intellectual context the Italian translation of the *Droit des gens* was published in 1781 and intensified the interest of Venetian newspapers in international politics and the correlation of war, neutrality and peace. Lodovico Antonio Loschi was particularly active in organising the circulation of books on the issue between Naples and Venice, sometimes by dint of contacts between Ferdinando Galiani and the Venetian senator Angelo Querini. A case in point was the *Memoria riguardante il sistema di pace e di guerra che le potenze straniere praticano con le reggenze di Berberia* (Messina: 1786) by Bartolomeo Forteguerri of Livorno,[19] which Loschi received from Galiani via Querini and immediately republished along with other writings on Venice's position in the international system.[20]

It is from this perspective that a reconstruction of the events and activities through which the *Droit des gens* was received and used in the Italian small states system to aid constitutional and policy reforms assumes most of its relevance. Until now Italian historiography has paid scant attention to Vattel's work. Recent research has already shed some light on the wide-ranging debate in the peninsula during the late Enlightenment, in which the *Droit des gens* was used as a guide for a new science of politics to which a significant number of attempted translations into Italian bear witness.[21] The recovery of these translations—which were made, independently of one another, throughout the peninsula, from Venice and Florence to Naples and Sicily—opens a window onto cultural exchanges that took place in the late eighteenth and early nineteenth centuries. It also invites questions that require detailed answers. Why did the ruling and educated classes, who were able to read Vattel's work in French, feel the need to produce these Italian translations? And why, following the first edition of 1758, was there this need, twenty years later, in the late 1770s and early 1780s, to continue producing them? Examining the Italian translations of Vattel's work in a wider context reveals that the translation strategies of the Enlightenment responded to the need for a new political vocabulary and a common European lexicon.

Vattel's work began to circulate very early on in Savoyard Piedmont, thanks in part to its proximity to Switzerland. Sales of the *Droit des gens* took place in Geneva and Turin from 1759 by means of a circuit that later stretched as far as the Naples bookseller Gravier, who in turn was in contact with the Gosse printers.[22] Fifteen years later booksellers in Turin still continued to acquire Vattel's "excellent treatise"[23] from the Societé typographique de Neuchâtel.

In Piedmont the importance of Vattel's work was not limited to the realms of theory but had an influence on actual government practice. The *Droit des gens* became crucial to how diplomatic relations were conducted between states, and in particular between Turin and Geneva.[24] The work, together with other classics on international law, was to be found in the library of Luigi Malabaila di Canale, ambassador to Vienna.[25] In 1772 Carlo Ignazio Montagnini (1730–1790), later secretary of legation in Vienna, borrowed from Vattel when replying to the demands of the French ambassador to Turin, who wished to have the same privileges that his predecessors had enjoyed, specifically that of exemption from taxes on meat. Vattel's authority was used to argue that such diplomatic privileges depended not on a right but on the simple will of the sovereign, on his "civilité" (graciousness) and on the discretion considered appropriate for representatives of other states.[26]

A later but very significant testimony was given by Gian Francesco Galeani Napione (1748–1830), who held important positions in the administration of Savoy finances. In 1799 he recollected how decades earlier many private tutors had utilised Grotius, Pufendorf, Bielfeld and Vattel when teaching public law, a use that, seen retrospectively in the light of revolutionary events, must have been considered dangerous for the government, especially when there was no control over the teachers in question.[27]

FROM MARITIME TO INTERNAL POLICY

During the eighteenth century, the states of the Italian peninsula still confronted the model in which restricted territorial size and virtue were coupled.[28] It is within this process—the shift from a classical political model to self-aware modernity—that the influence of and interest in Vattel's work in the 1770s and 1780s must be placed. In Tuscany signs of the use of Vattel in government practice can be seen when the concept of territorial waters that gave states sovereign rights over the area of the sea within the distance that could be covered by a cannon shot was accepted even for a neutral power like the Grand Duchy.[29] These principles were translated into practical rules thanks to the measures taken by the Grand Duke to establish a ban, on 1 August 1778, prohibiting hostile acts between enemy nations in the port of Livorno and the adjacent waters within the range of a cannon, with the obligation on all nations to abstain from any plunder, pursuit, inspection or act of violence that might contravene the principle of equality between nations and the neutrality guaranteed by Tuscany.[30]

The context in which the *Droit des gens* came to be used and adapted in the Grand Duchy of Tuscany during the late eighteenth century differed profoundly from the one in which it had been written. The book was pressed into the service of the process of constitutionalisation and state transformation carried out in the wake of the American Revolution, events that have up to now been little explored because of the scant attention paid to Vattel's work by Italian historiography. Closer evidence of how Vattel was used in the Duchy in the years immediately following the first edition of 1758 is, however, somewhat scarce and little can be gleaned from data on the Tuscan book trade, which, in contrast to other parts of the peninsula, does not seem to have sold significant copies of the work.[31] Some references to readings of the *Droit des gens* can be found here and there in contemporary literature, which often carried Vattel's name in the authoritative genealogy of writers on natural law running from Locke and Grotius through Hobbes and Cumberland and ending with the Swiss author. In one instance, during the funeral eulogy held in Florence Cathedral in 1765 in honour of Francesco Stefano di Lorena, Vattel was referred to as the interpreter of the "holy law of nature, daughter of the Eternal Mind, spirit of the world."[32]

Vattel's work was also employed in Tuscany in support of the Grand Duchy's maritime and international policy, above all in relation to the matter of the legal position of the free port of Livorno, since it contained sections that helped to define the concepts of territorial waters and neutrality. These issues had already been addressed in 1756 by Giuseppe Maria Buondelmonti in his brief *Ragionamento sul diritto della guerra giusta*, presented to the Accademia della Crusca and published two years before Vattel's work.[33] Nevertheless, the *Droit des gens* was certainly known to Giovanni Maria Lampredi, professor of public law at the University of Pisa from 1769, who in 1761 used it in his *De licentia in hostem* and then in *Juris publici universali, sive iuris naturae et gentium theoremata* (1776–1778) and *Del commercio dei popoli neutrali in tempo di guerra*

(1788).[34] Lampredi cited Vattel very little in his *Theoremata*, but interpreters have not hesitated to assert his reliance on Vattel's theories, especially those of the *Droit des gens* relating to trade carried out by neutral states in times of war.[35] There are well-founded reasons to argue that the law theorised in the 1770s which stated that a neutral power like Tuscany could exercise its maritime sovereignty over the part of the sea that could be covered by a cannon shot was in fact based on the work of Vattel and Lampredi.[36]

In the 1770s Vattel's work was frequently used to make clear the context of the international relations in which the Grand Duchy of Tuscany found itself, above all from the point of view of maritime competition. In the oration held in 1770 at the meeting of the Knights of Saint Stephen by the Florentine Marco Covoni Girolami (1742–1824), later a member of the senate, the main concern was the threat posed by southern African corsairs to the freedom of the seas, commerce and navigation. The author used Vattel to explain that a nation's strength on the international stage, even that of a small one like Tuscany, was reinforced by the fact that the nation itself was the expression of the authority of both its governor and of the individual interests ("des particuliers") that it was able to represent.[37]

In contrast, almost nothing is known of how Vattel may have been used in the debate on the internal reforms set in motion by the Grand Duke in the mid-1770s. This was an extensive process which encompassed various aspects of economic, social and institutional life, and which prepared the way for an undertaking that would radically change the legacy inherited from the Lorraine-era Regency: it was an ambitious plan in constant evolution destined to transform the regional state of Tuscany into a modern one. This made it necessary to have a face-to-face confrontation between the Grand Ducal authority and the local communities with the manifest intention of overhauling the entire architecture of public and administrative powers and the role of intermediary bodies, which had been fixed by medieval and Renaissance tradition.

Peter Leopold's project clearly demonstrated that in order to tackle the crisis of the *ancien régime* it was no longer enough to institute partial reforms, but rather it was vital to take direct action on the state's constitutional composition. His plan indicated that all of this could be achieved through renewed collaboration between the ancient communities and the intermediary powers. The objective was to bring in a model of a constitution of classes ("costituzione per ceti") established through clear agreement, to replace the dialectic between centralisation and decentralisation that characterised other European arrangements.[38] The project was thus based on the recognition of the role of institutions and local communities, on the duty of the peripheral government towards local agencies, and on the network of judicial administrations which would enable all parts of society to participate in the programme of reforms.[39]

The project hinged on the adherence to a Physiocratic form of economic policy that enabled the role and function of the communities to be enhanced while leaving the organisation of the territory almost intact, limiting itself in fact to rationalising and optimising the existing structures. The decisive phase came in March 1779 when Peter Leopold began to reflect on the possibility of drafting a true and proper constitution for Tuscany. The moment was particularly delicate because it coincided with the aftermath of the American Revolution and the stiffening of the policies of his brother Joseph II of Habsburg, who was becoming increasingly despotic.[40] According to the interpreters, Peter Leopold's work proceeded more or less autonomously, at least initially, partly because his family experience had taught him that in Tuscany reformist propositions tended to succumb to conservative pressure.[41] Indeed, it was in the years 1778 to 1780 that he reordered all the early thoughts and ideas that had accumulated in his mind.

TUSCANY AS A CONSTITUTIONAL LABORATORY

In the spring of 1779 Peter Leopold thus came to set out his first *Idee sopra il progetto della creazione degli Stati* (ideas on the project for the creation of the states).[42] This was a series of notes that he had developed step by step as the debate on the reform of the territorial administration progressed, and it was this that he used, soon afterwards, for the first draft for a planned constitution, the *Primo disteso ed idee di S. A. R. sopra la formazione degli Stati e nuova costituzione pubblica.*[43] This was the moment in which all the results of the study and experience accumulated in earlier years were brought together in a single document. The text demonstrates his commitment to the rereading of the great tradition of European natural law which from Erasmus through Pufendorf, Hobbes, Grotius, Locke and Wolff had reached Montesquieu, Rousseau and the Physiocrats, amongst whom was Turgot.[44] The echo of the teachings of Locke and natural law was clear, linked to an ever-growing concern to take into account all the needs of Tuscan society. As has been amply demonstrated, the assumption was that every future modernising endeavour in Tuscany would have to be based on a political pact, on a kind of "management contract," which the prince would make with the "well composed society" that had arisen from the ruins of the old organisation of the communities. Also clear was the discontinuity that Peter Leopold imagined between this contract, which he termed a "management," in other words an agreement in which the prince had to limit himself to leading his people, in keeping with the classic tradition of the political pact situated at the foundation of society, as derived from the medieval tradition and based on a compromise between the classes. The new society that concluded this contract with the prince had to be different from the old one based on the "unjust division" into classes or orders. From this point of view, Peter Leopold had very clear ideas and was determined to distinguish himself from the "various authors that have written different books on the formation and constitution of different states."[45]

There has been great insistence on the fact that the ideas regarding the state and the relationship between prince and subjects that appeared in Tuscany depended to a great degree on the natural law conception of the state's fundamental law, which Frederick II of Prussia made well-known.[46] Nevertheless, apart from these suggestions, it has been difficult for scholars to identify the sources that Peter Leopold drew on, apart from the Pennsylvania Constitution to which he referred openly.[47] Moreover, it has been deemed improbable that, at least in this first phase, he had help from collaborators in Tuscany, where there would not have been a great deal of awareness of the sources and ideas that inspired those pages.

"A Well Organised Society"

Some historians have also opined that Peter Leopold's project has echoes of the *Droit des gens*, especially where his concept of fundamental law seemed to evoke older pacts that were typical of the Germanic world,[48] but no in-depth study has been made of this hypothesis. However, a more direct comparison with the earliest versions of the project, not yet influenced by subsequent revisions, makes it possible to discern Vattel's influence with some degree of precision. In particular, reference was made to the first book of the *Droit des gens*, entirely devoted to the constitutional architecture of the state, in which one finds the first theorisations of the difference between constitution and fundamental law, and between constituent power and legislative power.[49] This is also the book in which it is explained that while "la Nation est en plein droit de former elle-même la Constitution" (the nation is *ipso jure* the founder of the constitution), exercising a right which is inherent and cannot be delegated, the legislative authority—as a constituted power—could only be entrusted "au Prince, ou à une Assemblée, ou à cette Assemblée et au Prince conjointement" ("to the prince, or to an assembly or to this assembly and prince together").

In the first lines of the *Disteso* Peter Leopold openly declared that for him the state was a "society of men grouped together under a government" and in this sense there was clear agreement with Vattel, according to whom "nations or states are the political corps, societies of men united together to gain their health and advantage."[50] On the other hand, Peter Leopold continued:

> In a well organised society all and any member of the society must have an equal right to happiness, wellbeing, security and property, which consist in the free, tranquil and secure enjoyment and dominion of their own property and as a consequence also in being able to guard this and in influencing legislation, which has to constrain everybody.[51]

In these lines there was a recognisable consonance with the *incipit* of the Pennsylvania Constitution and with the Declaration of Independence of the American colonies. Yet they also harkened back, in this case, to the words of Vattel (who had been an inspiration to the American revolutionaries) where he affirmed that: "the Nations are free and independent of one another, because men are naturally free and independent; the second general law of their society is that each nation should be left to the peaceable enjoyment of this freedom, which is in its nature."[52]

However, it was Peter Leopold's shift from the image of the "states" towards the formulation of a "management contract" which presupposed the idea of a constitution no longer only deducible from a pre-established order but conceived, following Vattel's example, as an expression of the freedom of a nation that was able to establish the rules of its own political existence.[53] Then, according to Peter Leopold:

> by fundamental law of the state one means the contract with which authority has been accorded to the social states, and which limits and prescribes the duties and obligations, the faculty and authority of subjects or of the members that make up society, towards their first magistrate or towards he who

is charged with executive power, be it a sovereign or a magistrate, and their duties and obligations towards the subjects.[54]

In this case, too, the agreement with the *Droit des gens* is notable, since Vattel had explained that:

> the fundamental rule which determines the way in which the public authority must be exercised is that which forms the constitution of the state. In it one sees the forms on which the nation acts in its qualities as a political body; how and by whom the people must be governed, what are the rights and the duties of those who govern.[55]

For Peter Leopold the greater complication was that of identifying who he was dealing with. This could no longer be the sum of the communities or of the classes, but rather had to be—thanks to a curious word as yet not carefully investigated by scholars—"the universality" of the Grand Duchy.[56] This expression, which for the Grand Duke summed up his new idea of Tuscan society resulting from the process of reforms, appears once more to echo Vattel, where he defined the universal society as men bound to one another by natural reciprocal duties: "The universal society of human beings is an institution of nature itself, that is to say a necessary consequence of the nature of man; all men, in whatever state they may be in, are obliged to cultivate and fulfil their duties. They cannot be dispensed by any convention, for any particular association."[57]

In these first drafts of Peter Leopold's project, the question of how this "universal society" would express itself politically still remained open. In the opinion of the Grand Duke, in fact,

> in the countries in which there exists a division in social states, these are divided into clergy, nobility, and also in some places the cities, which form a state by themselves; but this is all a mistake; in no state or society can there be more than one social state, which is then divided into two classes, namely those who are owners and those who are not.[58]

Thus the solution seemed to him to be that of establishing "a corps of public representation" that brought together the "general deputies" who had been elected by all the provincial delegates to "run the affairs of the whole country."[59] Even here there is an idea of representation not dissimilar to that of Vattel: "Essentially it belongs to the Society to make laws on the way it claims to be governed, and on the conduct of the citizens: this can be called legislative power. The nation may entrust the exercise to the prince, or to an assembly, or to this assembly and prince together."[60]

The key part of this constitutional architecture therefore had to be that "management contract" whose function had actually been explained by Vattel through the definition of the prince as a "leader":

> The political society is a moral person in that he has understanding and will, which he uses to conduct his affairs, and is capable of obligations and rights. So when he confers sovereignty on someone, he places in him his understanding and will, he transfers his obligations and rights to him, as they relate to the administration of the state, to the exercise of the public authority; and the leader of the state, the sovereign, restores to the subject or resident the obligations and rights relating to government, the moral person is found in him who, without completely ceasing to exist in the nation, henceforth never acts other than in him and for him. This is the origin of the representative character attributed to the sovereign.[61]

We know that from the last months of 1781 onwards Peter Leopold began to discuss his project with a larger circle of collaborators, and the constitutional plan thereafter underwent many adaptations and modifications, often the fruit of compromises between the Grand Duke's aspirations and the reality presented to him by government officials. The subsequent versions of the constitutional project, widely influenced by the interventions of advisers, including the minister Francesco Maria Gianni, lost a large part of the project's original force, conforming to a canon of natural law and Physiocracy, which made the impact of Vattel less noticeable.[62]

NOTES

1. Emer de Vattel, *Le droit des gens ou principes de la loi naturelle appliqués à la conduite et aux affaires des nations et des souverains. Nouvelle édition augmentée* (Neuchâtel: Société Typographique, 1773).

2. Emer de Vattel, *Le droit des gens ou principes de la loi naturelle appliquée à la conduite des affaires des nations et des souverains. Nouvelle edition* (Neuchâtel: De l'Imprimerie de la Société Typographique, 1777), vol. 1, xviii.

3. Alessandro Verri, "Di alcuni sistemi del pubblico diritto," *Il Caffè* 32 (1766); reprinted in *Il Caffè 1764–1766*, ed. Gianni Francioni and Sergio Romagnoli, vol. 2 (Turin: Bollati Boringhieri, 1998), 736–739.

4. Antonio Genovesi, *Delle lezioni di commercio o sia di economia civile, con Elementi del commercio*, ed. Maria Luisa Perna (Naples: Istituto italiano per gli studi filosofici, 2005), 291 (part 1, ch. 1, § 31).

5. Carlo Capra, *I progressi della ragione. Vita di Pietro Verri* (Bologna: il Mulino 2002), 172. The manuscript of the *Saggio* is in AV, ms. 375.2 (reference to Vattel on fol. 119).

6. Vattel, *Droit des gens*, book. I, ch. VIII, § 98, quoted by Pietro Verri, "Stato attuale del commercio di Milano," in *Opere*, vol. 2/2 *Scritti di economia, finanza e amministrazione*, ed. Giuseppe Bognetti, Angelo Moioli, Pier Luigi Porta and Giovanna Tonelli (Rome: Edizioni di Storia e Letteratura, 2006), 203.

7. Vattel, *Droit des gens*, book I, ch. XIX, § 223, ch. VI, § 73 and ch. XIII, § 166, quoted by Verri, "Stato attuale," 242, 244, 246.

8. Vattel, *Droit des gens*, book. I, ch. VIII, § 97 and ch. 21, § 260, quoted by Verri, "Stato attuale," 263, 705.

9. Vattel, *Droit des gens*, book. I, ch. XIV, § 183 and 185, quoted by Pietro Verri, "Sul tributo del sale nello Stato di Milano" in *Opere*, vol. 2/2, 695.

10. Verri's report of 11 February 1770 in AV, nr. 407, fol. 420, quoted by Capra, *I progressi della ragione*, 311.

11. "Annotazioni fatte al libro intitolato Meditazioni sulla economia politica con alcune brevi osservazioni," manuscript in AV, nr. 321, fol. 14. See also Franco Venturi, *Settecento riformatore*, V/1, *L'Italia dei Lumi* (Turin: Einaudi 1987), 552.

12. Vattel, *Droit des gens*, book I, ch. XIV, § 171, quoted by Cesare Beccaria, *Dei delitti e delle pene*, critical edition by Gianni Francioni, in *Opere*, vol. 1 (Milan: Mediobanca, 1984), § 11.

13. Gianni Francioni, "Nota al testo," in Beccaria, *Dei delitti e delle pene*, 215–368.

14. Alberto De Simoni, *Del furto e sua pena, trattato con alcune osservazioni generali in materia criminale* (Lugano: Agnelli, 1776), 13; Renato Pasta, "Beccaria tra giuristi e filosofi: aspetti della sua fortuna in Toscana e nell'Italia centrosettentrionale," in *Cesare Beccaria tra Milano e l'Europa*, ed. Sergio Romagnoli and Gian Domenico Pisapia (Milan-Rome: Cariplo-Laterza, 1990), 515.

15. Adriano Cavanna, *La codificazione penale in Italia. Le origini lombarde* (Milan: Giuffrè, 1987), 101.

16. Ezio Godoli, *Trieste* (Rome-Bari: Laterza, 1984), 87.

17. Giovanni Scarabello, "Il Settecento," in *La repubblica di Venezia in età moderna, dal 1517 alla fine della Repubblica*, ed. Gustavo Cozzi, Michael Knapton and Giovanni Scarabello (Turin: Utet, 1992), 565–567.

18. *Riflessioni di un francese sopra i motivi della guerra francese*, manuscript in FQS, translation of *Réflexions d'un Suisse sur les motifs de la guerre présente* (The Hague: Varon, 1757). The manuscript is currently being studied by Koen Stapelbroek, in advance of a critical edition.

19. ASM, Letterati, Loschi, nr. 37/14; Mirella Mafrici, *Il mezzogiorno d'Italia e il mare: problemi difensivi nel Settecento*, http://www.storiamediterranea. it/public/md1_dir/b703.pdf, 657–658.

20. ASM, Letterati, Loschi, busta 37/2, fol. 34rv: "Memoire presenté par le Résident de Venise à L.H.P. et delivré à Vienne par son Ambassadeur aux Ministres étrangers"; no. 58: "Extrait de la Conjurationdes espagnols contre la République de Venise en l'année MDCXVIII par mr l'abbé de S. Rent."

21. More details in Antonio Trampus, "The circulation of Vattel's Droit des gens in Italy: the doctrinal and practical model of government," in *War, Trade and Neutrality: Europe and the Mediterranean in the seventeenth and eighteenth centuries*, ed. Antonella Alimento (Milan: FrancoAngeli, 2011), 217–232.

22. Lodovica Braida, *Il commercio delle idee. Editoria e circolazione del libro nella Torino del Settecento* (Florence: Olschki, 1995), 259.

23. Braida, *Il commercio delle idee*, 276.

24. Dino Carpanetto, *Divisi dalla fede. Frontiere religiose, modelli politici, identità storiche nelle relazioni fra Torino e Ginevra (XVII-XVIII secolo)* (Turin: Utet, 2009), 248–249.

25. Ada Ruata, *Luigi Malabaila di Canale. Riflessi della cultura illuministica in un diplomatico piemontese* (Turin: Deputazione subalpina di storia patria, 1968), 134.

26. AST, sez. I, Casa Reale, Cerimoniale ambasciatori ed., inviati, mazzo 1, n. 14/II: "Mémoire touchant les Franchises des Ministres Entragers. Dressé par Mons.r Montagnini Chargé des affaires de S.M. auprès de la Cour de Vienne." For this context Donatella Balani, *Il vicario tra città e stato. Ordine pubblico e annona nella Torino del Settecento* (Turin: Deputazione subalpina di storia patria, 1987), 230–232; Daniela Frigo, *Principi ambasciatori e jus gentium. L'amministrazione della politica estera nel Piemonte del Settecento* (Rome: Bulzoni, 1991), 248, 266.

27. Gian Francesco Galeani Napione, *Del modo di riordinare la Regia Università degli studi*, ed. Paola Bianchi (Turin: Deputazione Subalpina di Storia patria, 1993), 144.

28. For a general overview see Maurizio Bazzoli, *Il piccolo Stato nell'età moderna. Studi su un concetto della politica internazionale tra XVI e XVIII*

secolo (Milan: Jaca Book, 1990); Carlo Capra, "The Italian States in Early Modern Period," in *The Rise of the Fiscal State in Europe c. 1200–1815*, ed. Richard Bonney (Oxford: Oxford University Press, 1999), 417–439; Matthias Maass, *Small states in world politics: The story of small states survival 1648–2016* (Manchester: Manchester University Press, 2017), ch. 3.

29. Vattel, *Droit des gens*, book II, ch. XX, § 289.
30. Georg Friedrich Martens, *Recueil des traités d'alliance, de paix, de trève* etc., vol. 3 (Göttingen: Dieterich, 1818), 24; Tullio Scovazzi, "The Evolution of International Law of the Sea: New Issues, New Challenges," in *Recueil des cours. Collected courses of the Hague Academy of International Law*, ed. Académie de droit international (The Hague: Nijhoff, 2001), 39–244.
31. See the sales records of the book dealers Luchtman of Leiden in UA, "Koninklijke vereniging voor het boekenvak, Luchtmans archief boekverkopers boeken 1697–1803."
32. Giovanni Giorgio degli Alberti in *Raccolta di tutto ciò che si è fin qui pubblicato in Livorno e altrove in morte di... Francesco I*, vol. 2 (Livorno: Strambi, 1765), 181; Eric W. Cochrane, *Tradition and Enlightenment in the Tuscan Academies 1690–1809* (Rome: Edizioni di Storia e letteratura, 1961), 232
33. Giuseppe Maria Buondelmonti, *Ragionamento sul diritto della guerra giusta letto nell'Accademia della Crusca* (Florence: Bonducci, 1757); Furio Diaz, "Buondelmonti Giuseppe Maria," in *Dizionario biografico degli italiani*, vol. 15 (Rome: Istituto della Enciclopedia Italiana, 1972), 212–215.
34. Giovanni Maria Lampredi, *De licentia in hostem liber singularis in quo Samuelis Cocceii sententia de infinita licentia in hostem exponitur et confutatur* (Florence: Excudebant Imperiales Typographi, 1761), 10; Giuseppe Maria Lampredi, *Juris publici universalis sive juris naturae et gentium theoremata* (Livorno: n.p., 1776–1778), 207; Giovani Maria Lampredi, *Del commercio dei popoli neutrali in tempo di guerra* (Milan: n.p. 1788). On Lampredi see Paolo Comanducci, *Settecento conservatore. Lampredi e il diritto naturale* (Milan: Giuffrè, 1981); Fabrizio Vannini, "Lampredi Giovanni Maria," in *Dizionario biografico degli italiani*, vol. 63 (Rome: Istituto della Enciclopedia Italiana, 2004), 259–262.
35. Comanducci, *Settecento conservatore*, 267–269.
36. Vattel, *Droit des gens*, book II, ch. XX, § 289.
37. Marco Covoni, *Orazione recitata nel solenne Capitolo de' Cavalieri di S. Stefano* (Florence: Allegrini, 1770), vi (with reference to Vattel, *Droit des gens*, book I, ch. XV, § 18).
38. Bernardo Sordi, *L'amministrazione illuminata. Riforma delle comunità e progetti di costituzione nella Toscana leopoldina* (Milan: Giuffrè, 1991);

Luca Mannori-Bernardo Sordi, *Storia del diritto amministrativo* (Rome-Bari: Laterza, 2003), 194–195.

39. Sordi, *L'amministrazione illuminata*, 14–15. The Austrian documents of Peter Leopold on the Tuscan constitution are now available on the website of the *Centro Interuniversitario di ricerca sulla storia delle città Toscane* (CIRCIT), www.circit.it

40. Adam Wandruska, *Leopold II* (Vienna: Herold, 1963); Italian translation: *Pietro Leopoldo. Un grande riformatore* (Florence: Vallecchi, 1968), 390. See also Peter Leopold, *Relazioni sul governo della Toscana*, vol. 1–3, ed. A. Silvestrini (Florence: Olschki 1969–1974).

41. Furio Diaz, *Francesco Maria Gianni. Dalla burocrazia alla politica sotto Pietro Leopoldo di Toscana* (Milan-Naples: Ricciardi, 1966), 282.

42. ASF, Segreteria di Gabinetto, filza 167, ins. 21 "Idee sopra il progetto."

43. ASF, Segreteria di Gabinetto. Appendice, filza 10, ins. 1

44. Joseph Zimmermann, *Das Verfassungsproject des Grossherzogs Peter Leopold von Toskana* (Heidelberg: Carl Winter, 1901), 83; Adam Wandruszka, "Joseph II und das Verfassungsproject Leopolds II," *Historische Zeitschrift* 190 (1960); Wandruszka, *Pietro Leopoldo*, 387 and 393.

45. ASF, Segreteria di Gabinetto, filza 167, ins. 21 "Idee sopra il progetto," c.5v. and 19v.

46. Sordi, *L'amministrazione illuminata*, 317–328.

47. This documents are in HHStA, Familienarchiv, Sammelbände, Kart. 13, "Fogli da aggiungersi alli Stati generali," ins. 10–11, 12, "Constitution de la république de Pennsylvanie."

48. Sordi, *L'amministrazione illuminata*.

49. Vattel, *Droit des gens*, book I, ch. XXXI and XXXIV.

50. Vattel, *Droit des gens*, Preliminaries.

51. "In a well-ordered society all and any members of the same society must have an equal right to happiness, wellbeing, security and property, which consist in the free, tranquil and secure enjoyment and dominion over their own things and as a consequence also in being able to be able to watch over it and to influence the legislation, which must be the task of everyone," ASF, Segreteria di Gabinetto, Appendice, F. 10, ins. 1 cc. 5r-v., quoted in Diaz, *Francesco Maria Gianni*, 287.

52. Vattel, *Droit des gens*, book I, ch. II, § 15.

53. Sordi, *L'amministrazione illuminata*, 326–327; see also Marina Valensise, "La constitution française," in *The French Revolution and the Creation of Modern Political Culture*, vol. 1, *The Political Culture of the Old Regime*, ed. Keith M. Baker (Oxford: Oxford University Press, 1987), 445–446.

54. "By 'fundamental law of the state' is meant the contract with which authority has been granted to the social states, and which limits and prescribes the duties and obligations, the faculties and authorities of the subjects or mem-

bers that make up society towards their first magistrate or the one in charge of executive power, be it a sovereign or a magistrate, and the duties and obligations of the latter to the subjects," *Idee sopra il progetto*, c. 11r.

55. Vattel, *Droit des gens*, book I, ch. III, § 27.
56. "Idee sopra il progetto," manuscript, c. 10r.
57. Vattel, *Droit des gens*, Preliminaries.
58. "In countries where there is a division into social states, these are divided into ecclesiastics, nobility, and even in some places cities, which form a state in themselves. But this is completely wrong: in no state or society can there be any more than one social state, at most divided into two classes, that is, those who are landowners and those who are not," "Idee sopra il progetto," manuscript, c. 19v.
59. Idee sopra il progetto," manuscript, c. 19r.
60. Vattel, *Droit des gens*, book I, ch. III, § 34.
61. Vattel, *Droit des gens*, book I, ch. IV, § 40.
62. See the annotation by Gianni in ASF, *Editto diviso in tre parti*, Proemio, cc. 2rv

BIBLIOGRAPHY

Most of the quotations of Vattel's *Droit des gens* come from the widely available English-language edition of 2008, which includes an introduction by Béla Kapossy and Richard Whatmore (Indianapolis: Liberty Fund) and maintains the text and English title *The Law of Nations: Or, Principles of the Law of Nature Applied to the Conduct and Affairs of Nations and Sovereigns* of the 1797 London standard edition.

Contemporary translations of primary sources are listed under the names of their authors. Full manuscript sources are referenced only in the notes for reasons of space.

ARCHIVAL ABBREVIATIONS

ASF   Archivio di Stato di Firenze, Florence, Italy
ASM   Archivio di Stato di Modena, Modena, Italy
AST   Archivio di Stato di Torino, Turin, Italy
AV   Archivio Verri, Fondazione Raffaele Mattioli, Milan, Italy
FQS   Fondazione Querini Stampalia, Venice, Italy
HHStA   Haus-, Hof- und Staatsarchiv Wien, Vienna, Austria
UA   Universiteitsbibliotheek Amsterdam, Amsterdam, The Netherlands

Primary Sources

Printed Books

Alberti, Giovanni Giorgio degli, *Raccolta di tutto ciò che si è fin qui pubblicato in Livorno e altrove in morte di... Francesco I*, vol. 2 (Livorno: Strambi, 1765)

Beccaria, Cesare, *Dei delitti e delle pene*, critical edition by Gianni Francioni, in: *Opere*, vol. 1 (Milan: Mediobanca, 1984)

Covoni, Marco, *Orazione recitata nel solenne Capitolo de' Cavalieri di S. Stefano* (Florence: Allegrini, 1770)

De Simoni, Alberto, *Del furto e sua pena, trattato con alcune osservazioni generali in materia criminale* (Lugano: Agnelli, 1776)

Galeani Napione, Gian Francesco, *Del modo di riordinare la Regia Università degli studi*, ed. Paola Bianchi (Turin: Deputazione Subalpina di Storia patria, 1993)

Genovesi, Antonio, *Delle lezioni di commercio o sia di economia civile, con Elementi del commercio*, ed. Maria Luisa Perna (Naples: Istituto italiano per gli studi filosofici, 2005)

Lampredi, Giovanni Maria, *De licentia in hostem liber singularis in quo Samuelis Cocceii sententia de infinita licentia in hostem exponitur et confutatur* (Florence: Excudebant Imperiales Typographi, 1761)

———, *Del commercio dei popoli neutrali in tempo di guerra* (Milan: n.p. 1788)

———, *Juris publici universalis sive juris naturae et gentium theoremata* (Livourne: n.p., 1776–1778)

Martens, Georg Friedrich, *Recueil des traités d'alliance, de paix, de trève etc.*, vols. 3 (Gottingue: Dieterich, 1818)

Peter Leopold of Tuscany, *Relazioni sul governo della Toscana*, vols. 1–3, ed. A. Silvestrini (Florence: Olschki 1969–1974)

Vattel, Emer de, *Law of Nations: Or, Principles of the Law of Nature, Applied to The Conduct and Affairs of Nations and Sovereigns*, eds. Béla Kapossy and Richard Whatmore (Indianapolis: Liberty Fund 2008)

———, *Le droit des gens ou principes de la loi naturelle appliqués à la conduite et aux affaires des nations et des souverains. Nouvelle édition augmentée* (Neuchâtel: Société Typographique, 1773)

———, *Le droit des gens ou principes de la loi naturelle appliquée à la conduite des affaires des nations et des souverains. Nouvelle edition* (Neuchâtel: De l'Imprimerie de la Société Typographique, 1777)

Verri, Alessandro, "Di alcuni sistemi del pubblico diritto", *Il Caffè* 32, 1766; reprinted in *Il Caffè 1764–1766*, eds. Gianni Francioni and Sergio Romagnoli, vol. 2 (Turin: Bollati Boringhieri, 1998), 736–739

Verri, Pietro, "Stato attuale del commercio di Milano," in *Opere*, vol. 2/2 *Scritti di economia, finanza e amministrazione*, eds. Giuseppe Bognetti, Angelo Moioli, Pier Luigi Porta and Giovanna Tonelli (Rome: Edizioni di Storia e Letteratura, 2006)

Secondary Sources

Balani, Donatella, *Il vicario tra città e stato. Ordine pubblico e annona nella Torino del Settecento* (Turin: Deputazione subalpina di storia patria, 1987)

Bazzoli, Maurizio, *Il piccolo Stato nell'età moderna: Studi su un concetto della politica internazionale tra XVI e XVIII secolo* (Milan: Jaca Book, 1990)

Braida, Lodovica, *Il commercio delle idee. Editoria e circolazione del libro nella Torino del Settecento* (Florence: Olschki, 1995)

Capra, Carlo, "The Italian States in Early Modern Period," in *The Rise of the Fiscal State in Europe c. 1200–1815*, ed. Richard Bonney (Oxford: Oxford University Press, 1999), 417–439

———, *I progressi della ragione. Vita di Pietro Verri* (Bologna: il Mulino 2002)

Carpanetto, Dino, *Divisi dalla fede. Frontiere religiose, modelli politici, identità storiche nelle relazioni tra Torino e Ginevra (XVII–XVIII secolo)* (Turin: Utet, 2009)

Cavanna, Adriano, *La codificazione penale in Italia. Le origini lombarde* (Milan: Giuffrè, 1987)

Cochrane, Eric W., *Tradition and Enlightenment in the Tuscan Academies 1690–1809* (Rome: Edizioni di Storia e letteratura, 1961)

Comanducci, Paolo, *Settecento conservatore. Lampredi e il diritto naturale* (Milan: Giuffrè, 1981)

Diaz, Furio, *Buondelmonti Giuseppe Maria*, in *Dizionario biografico degli italiani*, vol. 15 (Rome: Istituto della Enciclopedia Italiana, 1972), 212–215

———, *Francesco Maria Gianni. Dalla burocrazia alla politica sotto Pietro Leopoldo di Toscana* (Milan-Naples: Ricciardi, 1966)

Francioni, Gianni, "Nota al testo," in Cesare Beccaria, *Dei delitti e delle pene*, critical edition by Gianni Francioni, *Opere*, vol. 1 (Milan: Mediobanca, 1984), 215–368

Frigo, Daniela, *Principi ambasciatori e jus gentium. L'amministrazione della politica estera nel Piemonte del Settecento* (Rome: Bulzoni, 1991)

Godoli, Ezio, *Trieste* (Rome-Bari: Laterza, 1984)

Maass, Matthias, *Small states in world politics: The story of small states survival 1648–2016* (Manchester: Manchester University Press, 2017)

Mannori, Luca, and Sordi, Bernardo, *Storia del diritto amministrativo* (Rome-Bari: Laterza, 2003)

Pasta, Renato, "Beccaria tra giuristi e filosofi: aspetti della sua fortuna in Toscana e nell'Italia centrosettentrionale," in *Cesare Beccaria tra Milano e l'Europa*, eds. Sergio Romagnoli and Gian Domenico Pisapia (Milan-Rome: Cariplo-Laterza, 1990), 512–533

Ruata, Ada, *Luigi Malabaila di Canale. Riflessi della cultura illuministica in un diplomatico piemontese* (Turin: Deputazione subalpina di storia patria, 1968)

Scarabello, Giovanni, "Il Settecento," in *La repubblica di Venezia in età moderna, dal 1517 alla fine della Repubblica*, vol. 2, eds. Gustavo Cozzi, Michael Knapton and Giovanni Scarabello (Turin: Utet, 1992), 551–681

Scovazzi, Tullio, "The Evolution of International Law of the Sea: New Issues, New Challenges," in *Recueil des cours. Collected courses of the Hague Academy of International Law*, ed., Académie de droit international (The Hague: Nijhoff, 2001), 39–244

Sordi, Bernardo, *L'amministrazione illuminata. Riforma delle comunità e progetti di costituzione nella Toscana leopoldina* (Milan: Giuffrè, 1991)

Trampus, Antonio, "The circulation of Vattel's Droit des gens in Italy: the doctrinal and practical model of government," in *War, Trade and Neutrality: Europe and the Mediterranean in the seventeenth and eighteenth centuries*, ed. Antonella Alimento (Milan: FrancoAngeli, 2011), 217–232

Valensise, Marina, "La constitution française," in *The French Revolution and the Creation of Modern Political Culture*, vol. 1, *The Political Culture of the Old Regime*, ed. Keith M. Baker (Oxford: Oxford University Press, 1987), 445–446

Vannini, Fabrizio, "Lampredi Giovanni Maria," in *Dizionario biografico degli italiani*, vol. 63 (Rome: Istituto della Enciclopedia Italiana, 2004), 259–262

Venturi, Franco, *Settecento riformatore*, vol. 5/1, *L'Italia dei Lumi (1764–1790)* (Turin: Einaudi, 1987)

Wandruska, Adam, *Leopold II* (Vienna: Herold, 1963; Italian translation: *Pietro Leopoldo. Un grande riformatore*, Florence: Vallecchi, 1968)

———, "Joseph II und das Verfassungsproject Leopolds II," *Historische Zeitschrift* 190 (1960), 18–30

Zimmermann, Joseph, *Das Verfassungsproject des Grossherzogs Peter Leopold von Toskana* (Heidelberg: Carl Winter, 1901)

CHAPTER 6

The Lost Manuscript and the First Italian Translation of Vattel's *Droit des gens*

The second and more far-reaching period of success for Vattel's work in Italy came after the significantly enlarged and improved new French editions of 1773 and 1777, a period intertwined with events influenced by echoes of the American Revolution and the new phase of political debate that began in the 1780s.[1] The reasons for the interest in Vattel in Tuscany during the late eighteenth century thus seem relatively easy to understand. Yet quite unknown is the fact that, at the same time as Peter Leopold's first ideas on the constitution were being drafted, right there in Florence Giovanni Benedetto Brichieri Colombi, a functionary and man of letters close to the court of the Grand Duchy for family and professional reasons, was working hard on an Italian translation of the *Droit des gens*.

Brichieri Colombi's family came from Finale Ligure (which belonged to the Spanish Habsburgs until 1713) and had reached prestigious heights with Giovanni Bernardo Brichieri Colombi (1682–1751), father of Giovanni Benedetto. Giovanni Bernardo graduated in law at Pavia and began his legal career in Milan, before moving to Vienna in 1729 as advocate of the Marquisate of Finale with responsibility for defending its interests after the establishment of the free port of Genoa. After leaving that office in 1745, he found a post in Lorraine Tuscany and the following year

Archival Abbreviations
ASF   Archivio di Stato di Firenze, Florence, Italy
ASS   Archivio di Stato di Siena, Siena, Italy

© The Author(s) 2020
A. Trampus, *Emer de Vattel and the Politics of Good Government*,
https://doi.org/10.1007/978-3-030-48024-0_6

was appointed fiscal auditor.² The eldest of his seven sons, Giovanni Domenico (Finale, 1716–Florence, 1787), became a close collaborator of Peter Leopold, first as a fiscal auditor and then, from 1784, as president of the Buon Governo, the organ created with the specific duty of supervising the affairs of the Grand Duchy's police.³

The youngest son, who interests us more directly, was Giovanni Benedetto. Born in 1729, he was ordained a priest in 1754 and graduated in law at Pisa two years later, whereupon he was given the post of judge of the court commissioner. Becoming an auditor of the Rota of Siena in 1765, in 1777 he was appointed auditor of the supreme magistrate in Florence, the body considered at that time to be the court of the prince. Then, from 1782, he became the auditor of the Florentine Rota and in his final years devoted himself to administering the family estate and caring for his grandchildren before his death in 1802.⁴

The Brichieri Colombi family was therefore closely linked, from the 1740s, to the Lorraine dynasty and to the Tuscan reform experience, first through Francis Stephen and then through Peter Leopold.⁵

THE TUSCAN TRANSLATION AND ITS FUNCTION

In the Brichieri Colombi family library, preserved in the State Archive of Siena, there are several manuscripts by Giovanni Benedetto on historical and legal issues. These include the *Notizie che riguardano il ministero della città di Firenze raccolte in diversi tempi da Giovanni Benedetto Brichieri Colombi*; a collection of notices and documents dating back to 1768; various juridical and procedural notes; an Italian index for *La science du gouvernement* by Réal de Curban;⁶ some *Appunti storico-geografici sul Valdarno di Sopra, Chianti, Casentino, Mugello e La Castellina*; some *Mezzi da praticarsi per facilitare la propagazione del genere umano secondo i dettami di una ben regolata morale*; an Italian translation of the *De iure asylorum tractatus locupletissimus auctore Georgio Rittershusio* (1624) under the title *Trattato del diritto dell'asilo del dottore Giorgio Rittershusio di Sultzbach con note di vario genere del dottore Casimiro Pozzi fiorentino*; a *Commento alle Istituzioni giustinianee in quattro libri*; and a collection of *Privilegi ed altro concernenti la Nazione ebrea*.⁷

The works on juridical and literary arguments appear to confirm the full involvement of Giovanni Benedetto in a cultural context strongly aimed at conciliating the modernisation efforts promoted by the Lorrainese dynasty with the tradition of pluralism and civic and communal autonomy that had characterised the Grand Duchy since the Medicean era. This is also confirmed by the legal role he carried out in the 1760s, during his term of office as auditor in the Siena Rota. As has been noted by scholars, Giovanni Benedetto was personally convinced of the centuries-old practice whereby the constitutional nature resided in the numerous political pacts that in various ways tied and subjected the communities of the Grand Duchy to Florence. In the reports he wrote for important pronouncements of the Sienese Rota he used traditional classifications of political agreements (*foedera*). These he set out, beginning with the *Corpus iuris civilis* and the ancient Roman historians, and brought together in the *tria genera foederum*, a system made up of the distinction between the *foedus aequum*, *foedus impar* and *foedus dediticium*, in other words, between the different articulations of political conventions or stipulations between independent communities, recognised by the interpreter even when he was dealing with agreements of subjugation rather than proper joint conventions.[8] Moreover, in a report in support of a decision made by the Sienese Rota in 1769, Brichieri Colombi argued that this related to "principles as much of our law, as of the people" and that the structure of the *foedus aequum* was the most useful since it clarified the relationships between the Tuscan cities and Florence as the dominant nation that preserves in part the sovereignty of the people united to it, limiting it in certain ways.

These are words that document well the continuity between the territorial organisation and policy of the Grand Duchy in the late Medicean and Lorrainese periods and the approach chosen by Peter Leopold, based on the idea of a constitutional state that gave up constructing a centralised apparatus but instead chose the Physiocratic model so as to minimise central intervention and allow the communities to govern themselves.[9]

It should come as no surprise therefore that an Italian translation of Vattel's book under the title *Il Diritto delle Genti / o sia / I Principi della legge naturale / Applicati alla condotta e agli Affari / Delle Nazioni e de' Governi / Del Signor de Vattel / Opera / che conduce a scoprire li veri interessi / Delle Potenze / Tradotta dal francese / Nell'Italiano* was found among Giovanni Benedetto Brichieri Colombi's papers.[10] This is a manuscript which, thanks to a simple handwriting comparison, can be attributed with certainty to him. It is made up of thirty paper files, each containing ten unnumbered sheets of paper written on both sides, making a total of 300 pages, all filled with dense handwriting on a single column that left a white, three-centimetre margin on the left. The first two files contain the preface, files 3–14 contain the first book, and files 15–30 the second book. The handwriting is clear and sharp and only files 3–8 are written in another hand, though these bear corrections penned by Brichieri Colombi; these files are not so easy to read because of the acidity of the inks used. The translation stops at section 354, chapter XVIII of book II of Vattel's work, coinciding with the end of the first volume in print in the 1758 edition and those of the 1760s.

As mentioned, files 3–8 are written in another hand and include a number of corrections by Brichieri Colombi. In attempting to identify this writing it was observed that it matches one used in another manuscript preserved in the same archives, namely the Italian version of the treatise on asylum law by Georg Ritterhausen,[11] which is accompanied by a letter, dated 30 August 1771, in which Brichieri Colombi claimed to be working on a translation made by his "young scholar," which he later corrected. If the comparisons are accurate we can therefore deduce that folders 3–8 were either copied or translated directly by the same "young scholar," and were then revised by Brichieri Colombi. We might therefore be looking at, in this case, a translation test set by him for his young collaborator.

The version of the *Droit des gens* seems on the whole to contain no second thoughts or substantial corrections and often, as in the note to paragraph 166, the few thoughts added by the translator follow a phrase that had been crossed out, which makes one think that this might have been a translation executed at one go with hardly a pause. The notes, which Vattel had placed at the foot of the pages, were inserted directly in the text and in some rare cases in the margin. As far as the edition of the *Droit des gens* used by Brichieri Colombi is concerned, the inclusion of a summary of the various chapters leads us to assume that it was an edition of the 1760s.

Attempting to date the manuscript is problematical as there is no explicit indication of what this might be. It is nevertheless possible to advance certain hypotheses suggested by the context. In the Tuscan culture and in the reformist vision of Peter Leopold interest in the themes addressed by the *Droit des gens* coincided with Brichieri Colombi's time in Florence: he was called as an auditor to the capital of the Grand Duchy in 1777 and remained there until 1782. Thus if the translation of Vattel's work is to be linked to the new constitution project set in course by Peter Leopold, then it must have been made some time in those five years. This hypothesis can be supported by more specific clues which allow us to determine even more precisely the translation period. The manuscript contains several inserts that add a summary of the contents of individual chapters but were not present in Vattel's original text. These were written on different pages to those that compose the files of the translation and which are in fact recycled letters—using the space not covered by writing—sent from Livorno to Brichieri Colombi, one of which was dated 18 September 1778. It therefore follows that he was certainly working on the translation during either 1778 or 1779. As for the interruption of the work, it is noteworthy that the version concludes with the first volume of the *Droit des gens* and does not extend into the second. Why might he have decided to abandon the work? The decision does not seem to be linked to any lack of interest in the contents of the second volume, which dealt entirely with international law and international relations, in contrast to the first volume that relates in the main to the domestic nature of state constitutions. Indeed, the themes linked to the nature and interpretation of international treaties are already addressed in the final chapters of the first volume so if Brichieri Colombi had intended to concentrate attention on internal public law, he would not have translated the pages of the first volume which related to international public law. It is therefore logical to assume that the translation was interrupted after the first volume for reasons unconnected to the translator's intentions, which may perhaps be traced back to the publication, carried out in Venice from 1781, of a complete Italian version by Lodovico Antonio Loschi. If this was the sequence, then 1781 can be considered to have been the *terminus ad quem* for the Florentine translation, which can therefore be placed with reasonable certainty between the years 1778 and 1781.

The Translative Choices

The manuscript, as has been said, has only a limited number of corrections and thus appears to have been ready for publication. The translation is literal, while the stylistic register, the lexical choices and the syntactic construction appear to adhere to those used by Tuscans of the period. The translative choices made by Brichieri Colombi can be better understood by means of a comparison between the original French and the Florentine version with the near contemporary Italian translation, the only one to have been published, by Loschi. From this one gains the impression that on the whole the variations are almost always stylistic and that the translation by Brichieri Colombi is more legible and free of the linguistic flourishes often used by Loschi.[12] An interesting example can be found in the introduction (*Préliminaires*) in which Vattel had explained that the law of nations is not originally anything other than natural law applied to the conduct of nations. He added:

> But, as the application of a rule cannot be just and reasonable if it is not done in a manner suitable to the subject, one should not think that the Law of Nations are precisely and everywhere the same thing as the natural law, are close subjects, so that one only needs to substitute nations for individuals.[13]

This sentence, which is quite complex even in the original French, was translated by Loschi in a way that made it almost incomprehensible, replacing expressions such as *sujet* and *particulier* with simple expressions in the Italian, and which had no particular political or juridical significance. In contrast, Brichieri Colombi's work reveals much greater attention to the meaning and the intelligibility of the words for an Italian-speaking public. This intention is recognisable in his reconsideration of how to translate the word *sujet*, which became first *suddito*, then *soggetto* and later *suddito* again.[14] At the end of the sentence, however, he retained the expression *particulier/particolare* (replaced by Loschi with *privato*) to refer to individual people. This is a choice that is certainly more reasonable in relation to the concept of 'universal society' evoked by both Vattel and Peter Leopold in the initial constitutional manuscripts, and it seems to conform more to the lexicon used in the preparative constitutional documents of 1780–1782,[15] where the noun *particolare*, referring to individuals, still appeared regularly.

Another interesting passage concerns the reflection inspired by the Physiocrats on the duties of governors and on abuses in agriculture. Here Vattel underlined the need to protect and encourage farmers:

> Yet another abuse of farming is the contempt with which the farmer is treated. The middle class of the towns, the craftsmen and even the most servile citizens, regard the farmer with a disdainful eye, humiliate him and discourage him: they dare to despise a profession that feeds mankind and is the natural vocation of man.[16]

Vattel's lexicon, as we see, is well-pitched in contrasting *laboureur* to *bourgeois* and to *artisan*. In Loschi's version this lost much of its incisiveness because he opted for the generic Italian terms *lavoratore, semplice abitante* and *artefice*.[17] For Brichieri Colombi, who was perhaps more aware of the Physiocratic implications of this lexicon, the contrariety between *contadini* and *cittadini* was introduced, and he kept the word *artigiani*.[18] The diversity and greater lexical precision exhibited by Brichieri Colombi were probably fruit of his reading Physiocratic texts and of the attention that the Grand Duke paid to these themes.[19]

The difference of quality between the two translations is also confirmed in other parts of the text, for example in those relating to the organisation of justice and to judicial administration. This point is seen where Vattel reminded the prince of his duty to choose capable and qualified judges:

> There is no disadvantage in entrusting the judgement of a trial to a company of people with wisdom, integrity and enlightenment; on the contrary this is something that the Prince could do better; and he has fulfilled in this respect all the duties he owes to his people, when he gave them judges adorned with all the qualities needed by ministers of justice. He is left only to watch over their conduct, so that they do not relax their efforts.[20]

Vattel's phrase, in the second part, undoubtedly presented problems of interpretation, translation and legibility for the Italian public, which Loschi could not overcome, once more offering a literal and scarcely comprehensible translation.[21] Brichieri Colombi, on the other hand, sought to resolve these problems in order to be legible and comprehensible by dint of a simplification of the grammar and of the sentence, albeit this took several attempts, as we can see from his crossing out of certain words in the manuscript.[22]

Another comparison of these different translative choices comes from an examination of the paragraphs dedicated by Vattel to the system of judicial guarantees and, in particular, to the pages in which he proposes to establish—as a guarantee for citizens and out of respect for fundamental laws—a supreme court with the power to adjudicate in the last instance with irrevocable sentences, or sentences that at least could not be altered by the prince:

> Once this sovereign tribunal is established, the prince cannot influence its decisions, and in general he is absolutely obliged to guard and protect these forms of justice. Seeking to violate them is to fall into arbitrary domination, to which one can assume no nation has ever wished to subject itself.[23]

Once again, Loschi translated literally and thereby obscured the issue for Italians, for whom the idea of a supreme magistracy beyond the control of the prince was completely inconceivable.[24] Brichieri Colombi instead chose a more liberal and therefore more successful translation which emphasised the limits of the prince's will.[25]

The impression that Brichieri Colombi's manuscript was destined for publication comes from another factor, namely a handwritten sheet inserted at the end entitled "passages to correct." It included only two points intended, as far as can be seen, to be notes to the text. The first referred to the need to change book 1, ch. 5, § 67, which dealt with the canonical obstacles to marriages not approved by the Holy See; the second referred to book 1, ch. 7, § 79, which dealt with how estates belonging to the Church in Spain hindered agricultural development. Brichieri Colombi did not expand these points any further, probably because publication had been abandoned, but they are important in so far as they provide a greater understanding of the Italian context into which Vattel's work was to be inserted. The first point is in fact the same as the passage highlighted by Loschi, who believed it should be censured in the 1781 Venetian edition. The question related to the power of the Catholic Church to deny the sacrament of matrimony or to impede the divorce of a foreign prince even if it was dictated by reasons of state. Vattel had argued that this intervention should be rejected as it was tantamount to interference in the domestic affairs of an independent and sovereign nation, while Loschi commented that such a declaration was typical of a Protestant.[26] The other passage concerned a criticism of the Spanish monarchy. In Loschi's version the translator's censure had been brought to bear on an observation made by Vattel about the Spanish showing contempt for native Americans and therefore for a free nation.[27] However, in Brichieri Colombi's version, the censure was aimed at Vattel's passages which were considered critical of the policy of the Church of Spain for having made it difficult to develop agriculture.

Both these cases, which are perhaps comprehensible given a possible objection by the ecclesiastical censor, recapture the religious climate and the context in which, with the right care, the *Droit des gens* could have again served as a precious resource. Tuscany's Jansenist politics, in particular, would have benefited from this reasoning, and it is not unusual to find the names of the Brichieri Colombi family, especially that of Giovanni Domenico, in the correspondence of the bishop Scipione de Ricci, where he commented on the notification of 1 February 1780 with which, in executing Peter Leopold's orders to bring some moderation to the theatres, he ordered the exclusion of foreign and Tuscan comedians, prohibited the opening of theatres during Advent and Lent, and limited the use of masks during carnival.[28] So, as has been noted,[29] even Peter Leopold's religious reforms, based on a series of diocesan synods promoted by the prince himself and culminating in the Synod of Pistoia, would evoke Vattel's work among their theoretical and political presuppositions, emphasising the system of natural law and the firm defence of the common good.

The events of the Tuscan translation of the *Droit des gens* thus appear to confirm certain lines of inquiry suggested by recent studies on the legitimisation of political discourse and on the strategies of consensus that accompanied Peter Leopold's constitutional project. In fact, in the Grand Duchy of Tuscany during the years 1778–1780 there were a significant series of translations and projects for editions of the classics of seventeenth-century political culture, from Schmidt d'Avenstein's *Principes de la législation universelle* to De Lolme's *Constitution de l'Angleterre ou état du Gouvernement anglais* and to Burlamaqui's *Principes du droit naturel et politique*, the Tuscan version of which was started in 1772 by Fortunato Bartolomeo De Felice, not without experiencing serious difficulties with the censor.[30] These were initiatives which, alongside the translation of Vattel, certainly support the notion of a relationship between these conceptual emergencies and the constitutional outlook of Peter Leopold's government and the needs and plans of Grand Ducal communicative strategies. It should be noted that, within a general picture, the Tuscan case was probably not isolated: one need only recall the similar operation set in motion by progressive Venetian circles between 1780 and 1781, which was also accompanied by the translation of Vattel alongside one of Burlamaqui.[31]

Naples in the Time of the League of Armed Neutrality

While in Italy in the 1760s and the early 1770s the use of the *Droit des gens* appeared to be substantially limited to the section on theories of sovereignty, which were useful for the study of internal dynamics as a function of reforms and of a renewal of their constitutional structure, the utilisation of Vattel changed at the end of that period, becoming more focused on issues of international politics. This marked a turning point in Vattel's fortunes that was undoubtedly due to the great events of international politics caused primarily by the end of the Seven Years' War and the birth of the United States of America, which also coincided with a new edition of the *Droit des gens* published in 1777. This was also the era of intense diplomatic efforts that changed the international balance of power and accompanied the drafting of the treaties of 1778–1779 between the United States and France and between Spain and Portugal, and the Treaty of Teschen.

Ten years after the era of Genovesi and the *Il Caffè* group, and faced with the consequences of the American Revolution, many of the protagonists in those debates now occupied government positions in Italy and were busy directing the activities of princes and republican governments. This throws light on the new season in the success of Vattel's work in the Italian peninsula as well as the more frequent publication of a series of works that took a closer look at the question of relations between states and nations, from the *Jus publicum* by Giovanni Maria Lampredi (1778) to the *Scienza della legislazione* by Filangieri (1780), and from the Italian translation of Vattel (1781) up to the essay by Ferdinando Galiani on *Dei doveri de' principi neutrali* of 1782.

The Neapolitan contest had apparently been favourable to the reception and discussion of the theories of the *Droit des gens* and natural law,[32] as the writings of Genovesi showed. Vattel was used by Michele de Jorio in Naples in 1781 for the drafting of a project for a maritime code and then, in the *Istruzioni particolari del Codice*, he openly availed himself of Vattel's preface to explain that the law was compulsory for nations, and political society, and individuals. According to Jorio, Vattel's work was invaluable because in it he had not only applied the ideas he had taken from Christian Wolff to the actual behaviour of nations and rulers, but had also demonstrated how the principles of natural law were applicable—in addition to the public law of individual nations and among nations—even to the arts, agriculture and trade. Moreover, De Jorio opined, the lucidity of the *Droit des gens* helped to identify the sources from which the rules of those highly important themes were derived.[33]

Ferdinando Galiani made a notable reference to Vattel in his *De' doveri dei principi neutrali* (1782): he acknowledged that Vattel had grasped the importance of the issue of neutrality yet he rebuked him for keeping the discourse in general terms without straying too far from Grotius, Wolff and Coccejus, and at times confusing the principles of justice with advice dictated by political prudence. According to Galiani, such an approach made it difficult to understand the chain of events.[34] However, he accepted and drew on Vattel's ideas when he defined the nation as a society of civilised men, irrespective of what form of government it had—monarchical, republican or mixed. What counted, indispensably, was that each nation was independent of the others and had the right to war, peace and the free and spontaneous entitlement to decide whether to remain neutral.[35] Far from belonging to the realm of utopian ideas and theories, Galiani's essay was firmly linked to the geographical and cultural context in which it was written. Furthermore, it was written against the background of the problems of piracy, pillaging and the lack of free trade in the Mediterranean, which prompted Catherine of Russia to promote the League of Armed Neutrality and to extend Russian influence in that area. Galiani thus used, criticised and developed Vattel's ideas on neutrality and the equality of nations at the very time when, confronted with the fresh dangers of international politics, the Kingdom of Naples had to decide whether to join Catherine's League.[36]

Neutrality and trade were the cornerstones of an idea of foreign policy that was based on a renewed and more modern conception of the principle of European stability. As Vattel himself wrote,[37] Europe formed a political system interwoven by the relations and interests of the nations that no longer seemed to be isolated or self-contained units but rather the necessary components of a system within which they interacted vigorously. In eighteenth-century Italian thought it was therefore often the "spirit of commerce"—to use the words of Antonio Genovesi—that pointed the attention of intellectuals and men of government towards Vattel's book and his theories on the constitutional state, on the equality of nations and on neutrality. As has been noted in the past, the Italian culture of the eighteenth century linked good government with the spirit of commerce, often by referring to Montesquieu, in the belief that the spirit of commerce brought with it principles similar to those of good government, such as moderation, order and tranquillity, and to the extent of directly linking trade with peace.[38]

The Italian scene was, however, always dominated by a strong sense of realism, which led not so much towards the enlightened ideal of universal peace as towards the closer objective of promoting the efforts of enlightened governments in ensuring the balance between the states through a new trans-national political system characterised by shared civil and political values and practices.[39] Vattel was therefore more appreciated as an interpreter and populariser of Wolff, in particular as a critic of an abstract cosmopolitical model and as an advocate of overcoming a vision of the balance of power based on natural conflict between nations in favour of the duty of mutual assistance.[40]

NOTES

1. See Michel Schlup, ed., *L'édition neuchâteloise au siècle des Lumières. La Société typographique de Neuchâtel (1769–1789)* (Neuchâtel: Bibliothèque Publique et Universitaire, 2002), 249, 250, 258.

2. Gabriele Turi, "Brichieri Colombi Giovanni Domenico," in *Dizionario biografico degli italiani*, vol. 14 (Rome: Istituto della Enciclopedia italiana, 1972), 229–232.

3. On Giovanni Domenico see Marcella Aglietti, *I governatori di Livorno dai Medici all'Unità d'Italia. Gli uomini, le istituzioni, la città* (Pisa: ETS, 2009), 167.

4. Erminio Jacona and Patrizia Turrini, *Le carte Brichieri Colombi. Inventario analitico* (Rome: Ministero per i beni e le attività culturali-Direzione generale per gli archivi, 2003), 14–21.

5. On Giovanni Bernardo and Giovanni Domenico see Marcello Verga, *Da "cittadini" a "nobili". Lotta politica e riforma delle istituzioni nella Toscana di Francesco Stefano* (Milan: Giuffrè, 1990), 150–151.

6. Gaspard Réal de Curban, *La science du gouvernement, ouvrage de morale, de droit et de politique*, 5 vols. (Aix-La Chapelle [Amsterdam-Paris]: n.p., 1761–1765).

7. ASS, Carte Brichieri Colombi, nrr. 72–80

8. Luca Mannori, *Il sovrano tutore. Pluralismo istituzionale e accentramento amministrativo nel principato dei Medici (sec. XVI-XVIII)* (Milan: Giuffrè, 1994), 4–142, 53.

9. Mannori, *Il sovrano tutore*, 456–457.

10. ASS, Carte Brichieri, nr. 75.

11. ASS, Carte Brichieri, nr. 78.

12. See Antonio Trampus, "Il ruolo del traduttore nel tardo Illuminismo: Lodovico Antonio Loschi e la traduzione italiana del *Droit des gens*," in *Il linguaggio del tardo Illuminismo*, ed., Antonio Trampus (Rome: Edizioni di Storia e Letteratura, 2009).

13. Vattel, *Droit des gens*, Preliminaries.

14. *Il diritto della natura e delle genti ovvero principii della legge naturale applicati alla condotta e agli affari delle nazioni e de' sovrani scritta nell'idioma francese dal Sig. di Vattel e recata nell'italiano da Lodovico Antonio Loschi* (Lyon [Venice]: n.p. [Vitto], 1781), vol. 1, 2.

15. See the explication by Peter Leopold of the term *particolare*, ASF, Segreteria di Gabinetto, filza 167, ins. 14–15, now available at http://www.circit.it/uploads/15_167_14-15bis.pdf

16. Vattel, *Droit des gens*, book 1, ch. 7, § 80.

17. *Il diritto della natura e delle genti*, ed. Loschi, 88.

18. *Il Diritto delle Genti o sia I Principi della legge naturale*, manuscript, unnumbered pages.

19. See also the Italian translation of Mirabeau's *Les devoirs: La scienza cioè i diritti e i doveri dell'uomo. Opera divisa in quattro parti che contengono 1. La vita naturale dell'uomo 2. La sua vita agricola 3. La sua vita sociale 4. La sua vita politica. Tradotta dalla prima edizione francese di Losanna dell'Anno 1773 da un accademico etrusco* (Florence: Cambiagi, 1774). About this text, see Antonio Trampus, *Storia del costituzionalismo italiano nell'età dei Lumi* (Rome-Bari: Laterza, 2009), 189–190.

20. Vattel, *Droit des gens*, book I, ch. XIII, § 163.

21. *Il diritto della natura e delle genti*, ed. Loschi, 174.

22. *Il Diritto delle Genti o sia I Principi della legge naturale*, manuscript by Brichieri. See book I, ch. XIII, § 163.

23. Vattel, *Droit des gens*, book I, ch. XIII.

24. *Il diritto della natura e delle genti*, ed. Loschi, 176.

25. *Il Diritto delle Genti o sia I Principi della legge naturale*, manuscript by Brichieri.

26. Trampus, "Il ruolo del traduttore nel tardo Illuminismo," 98–99.

27. Trampus, "Il ruolo del traduttore nel tardo Illuminismo," 100.

28. Codignola, Ernesto, ed., *Carteggi di giansenisti liguri*, vol 1 (Florence: Le Monnier, 1941), 365–366.

29. Mario Rosa, *Settecento religioso. Politica della ragione e religione del cuore* (Venice: Marsilio, 1999), 108.

30. Luigi Delia, "The Enlightenment, Encyclopedism and the Natural Rights of Men: The Case of the Code of Humanity (1778)," in *Thinking about the Enlightenment: Modernity and its Ramifications*, ed. Martin L. Davies (London-New York: Routledge, 2016) 75–76.

31. Elisabetta Fiocchi Malaspina, *L'eterno ritorno del Droit des gens di emer de Vattel (secc. XVIII–XIX). L'impatto sulla cultura giuridica in prospettiva globale* (Frankfurt: Max Planck Institute for Legal History, 2017). On the Italian translator of Burlamaqui, Count Crispi, see Niccolò Guasti, *L'esilio italiano dei gesuiti spagnoli. Identità, controllo sociale e pratiche culturali (1767–1798)* (Rome: Edizioni di Storia e Letteratura, 2006), 281–282.

32. Vittorio Conti, ed., *La ricezione di Grozio a Napoli nel Settecento* (Florence: Centro editoriale toscano, 2002).

33. Carlo Maria Moschetti, ed., *Il Codice marittimo del 1781 di Michele de Jorio per il regno di Napoli*, vol. 1 (Naples: Giannini 1979), 287, 1375–1376 (on maritime peace), 1325 (on neutrality).

34. Ferdinando Galiani, *De' doveri de' principi neutrali verso i principi guerreggianti e di questi verso i neutrali* ([Naples]: n.p., 1782), 5; on Galiani see Koen Stapelbroek, *Love, Self-deceit and Money* (Toronto: University of Toronto Press, 2008); on the relations between Lampredi and Galiani see Gianfranco Miglio, *La controversia sui limiti del commercio neutrale fra Giovanni Maria Lampredi e Ferdinando Galiani* (Milan: Ispi, 1942); Comanducci, *Settecento conservatore*.

35. Maurizio Bazzoli, *Il piccolo Stato nell'età moderna: Studi su un concetto della politica internazionale tra XVI e XVIII secolo* (Milan: Jaca Book, 1990), 62; Koen Stapelbroek, "Universal Society, Commerce and the Rights of Neutral Tride: Martin Hubner, Emer de Vattel and Ferdinando Galiani," *COLLeGIUM: Studies Across Disciplines in the Humanities and Social Sciences* 4 (2008): 63–89.

36. Gennaro Maria Monti, *La dottrina dell'abate F. Galiani sulla neutralità e l'adesione di Ferdinando IV alla Lega dei Neutri* (Milan: Ispi, 1942); Stephen C. Neff, *The rights and duties of neutrals: A general history* (Manchester: Manchester University Press, 2000).

37. Vattel, *Droit des gens*, book III, ch. III, § 47.

38. Paolo Silvestri, *Il liberalismo di Luigi Einaudi, o del Buongoverno* (Soveria Mannelli: Rubbettino, 2008), 55–56.

39. Silvia Maria Pizzetti, "La costruzione della pace e di una società internazionale nell'Europa moderna fra jus gentium e cosmopolitismo (secoli XVII–XVIII)," in *Con la ragione e col cuore. Studi dedicati a Carlo Capra*, ed. Stefano Levati and Marco Meriggi (Milan: FrancoAngeli, 2008), 213.

40. Pizzetti, "La costruzione della pace," 216–218.

BIBLIOGRAPHY

Most of the quotations of Vattel's *Droit des gens* come from the widely available English-language edition of 2008, which includes an introduction by Béla Kapossy and Richard Whatmore (Indianapolis: Liberty Fund) and maintains the text and English title *The Law of Nations: Or, Principles of the Law of Nature Applied to the Conduct and Affairs of Nations and Sovereigns* of the 1797 London standard edition.

Contemporary translations of primary sources are listed under the names of their authors. Full manuscript sources are referenced only in the notes for reasons of space.

Archival Abbreviations

ASF   Archivio di Stato di Firenze, Florence, Italy
ASS   Archivio di Stato di Siena, Siena, Italy

Primary Sources

Printed Books

Codignola, Ernesto, ed., *Carteggi di giansenisti liguri*, 3 vols. (Florence: Le Monnier, 1941–1942)

Galiani, Ferdinando, *De' doveri de' principi neutrali verso i principi guerreggianti e di questi verso i neutrali* ([Naples]: n.p., 1782)

Mirabeau, Victor de Ruqueti, *Les devoirs: La scienza cioè i diritti e i doveri dell'uomo. Opera divisa in quattro parti che contengono 1. La vita naturale dell'uomo 2. La sua vita agricola 3. La sua vita sociale 4. La sua vita politica. Tradotta dalla prima edizione francese di Losanna dell'Anno 1773 da un accademico etrusco* (Florence: Cambiagi, 1774)

Réal de Curban, Gaspard de, *La science du gouvernement, ouvrage de morale, de droit et de politique*, 5 vols. (Aix-La Chapelle [Amsterdam-Paris]: n.p., 1761–1765)

Vattel, Emer de, *Il diritto della natura e delle genti ovvero principii della legge naturale applicati alla condotta e agli affari delle nazioni e de' sovrani scritta nell'idioma francese dal Sig. di Vattel e recata nell'italiano da Lodovico Antonio Loschi* (Lyon [Venice]: n.p. [Vitto], 1781)

Vattel, Emer de, *Law of Nations: Or, Principles of the Law of Nature, Applied to The Conduct and Affairs of Nations and Sovereigns*, eds. Béla Kapossy and Richard Whatmore (Indianapolis: Liberty Fund 2008)

Secondary Sources

Aglietti, Marcella, *I governatori di Livorno dai Medici all'Unità d'Italia. Gli uomini, le istituzioni, la città* (Pisa: ETS, 2009)

Bazzoli, Maurizio, *Il piccolo Stato nell'età moderna: Studi su un concetto della politica internazionale tra XVI e XVIII secolo* (Milan: Jaca Book, 1990)

Conti, Vittorio, ed., *La ricezione di Grozio a Napoli nel Settecento* (Florence: Centro editoriale toscano, 2002)

Delia, Luigi, "The Enlightenment, Encyclopedism and the Natural Rights of Men: The Case of the Code of Humanity (1778)," in *Thinking about the Enlightenment: Modernity and its Ramifications*, ed. Martin L. Davies (London-New York: Routledge, 2016), 69–85

Fiocchi Malaspina, Elisabetta, *L'eterno ritorno del Droit des gens di Emer de Vattel (secc. XVIII–XIX). L'impatto sulla cultura giuridica in prospettiva globale* (Frankfurt: Max Planck Institute for Legal History, 2017)

Guasti, Niccolò, *L'esilio italiano dei gesuiti spagnoli. Identità, controllo sociale e pratiche culturali (1767–1798)* (Rome: Edizioni di Storia e Letteratura, 2006)

Jacona, Erminio and Turrini, Patrizia, *Le carte Brichieri Colombi. Inventario analitico* (Rome: Ministero per i beni e le attività culturali-Direzione generale per gli archivi, 2003)

Mannori, Luca, *Il sovrano tutore. Pluralismo istituzionale e accentramento amministrativo nel principato dei Medici (sec. XVI–XVIII)* (Milan: Giuffrè, 1994)

Miglio, Gianfranco, *La controversia sui limiti del commercio neutrale fra Giovanni Maria Lampredi e Ferdinando Galiani* (Milan: Ispi, 1942)

Monti, Gennaro Maria, *La dottrina dell'abate F. Galiani sulla neutralità e l'adesione di Ferdinando IV alla Lega dei Neutri* (Milan: Ispi, 1942)

Moschetti, Carlo Maria, ed., *Il Codice marittimo del 1781 di Michele de Jorio per il regno di Napoli*, 2 vols. (Naples: Giannini 1979)

Neff, Stephen C., *The rights and duties of neutrals: A general history* (Manchester: Manchester University Press, 2000)

Pizzetti, Silvia Maria, "La costruzione della pace e di una società internazionale nell'Europa moderna fra jus gentium e cosmopolitismo (secoli XVII–XVIII)," in *Con la ragione e col cuore. Studi dedicati a Carlo Capra*, eds. Stefano Levati and Marco Meriggi (Milan: Franco Angeli, 2008), 209–241

Rosa, Mario, *Settecento religioso. Politica della ragione e religione del cuore* (Venice: Marsilio, 1999)

Schlup, Michel, ed., *L'édition neuchâteloise au siècle des Lumières. La Société typographique de Neuchâtel (1769–1789)* (Neuchâtel: Bibliothèque Publique et Universitaire, 2002)

Silvestri, Paolo, *Il liberalismo di Luigi Einaudi, o del Buongoverno* (Soveria Mannelli: Rubbettino, 2008)

Stapelbroek, Koen, "Universal Society, Commerce and the Rights of Neutral Trade: Martin Hübner, Emer de Vattel and Ferdinando Galiani," in *Universalism in International Law and Political Philosophy*, ed. Petter Korkman and Virpi Mäkinen, *Collegium. Studies across Disciplines in the Humanities and Social Sciences* 4 (2008) (Helsinki: Helsinki Collegium for Advanced Studies, 2008), 63–89

Stapelbroek, Koen, *Love, Self-deceit and Money* (Toronto: University of Toronto Press, 2008b)

Trampus, Antonio, "Il ruolo del traduttore nel tardo Illuminismo: Lodovico Antonio Loschi e la traduzione italiana del *Droit des gens*," in *Il linguaggio del tardo Illuminismo*, ed. Antonio Trampus (Rome: Edizioni di Storia e Letteratura, 2009a)

Trampus, Antonio, *Storia del costituzionalismo italiano nell'età dei Lumi* (Rome-Bari: Laterza, 2009b)

Turi, Gariele, *Brichieri Colombi Giovanni Domenico*, in *Dizionario biografico degli italiani*, vol. 14 (Rome: Istituto della Enciclopedia italiana, 1972), 229–232

Verga, Marcello, *Da "cittadini" a "nobili". Lotta politica e riforma delle istituzioni nella Toscana di Francesco Stefano* (Milan: Giuffrè, 1990)

CHAPTER 7

The Consequences of the American Revolution: From Naples to Venice

In the 1780s Vattel's success, along with the fallout from the American Revolutionary War, caused the cultural circles in Italy to shift their attention towards his reflections on the structure of society, and on the characteristics of nations and the rights of the people who composed them. In this sense the influence of the *Droit des gens* also extended to Gaetano Filangieri's *Scienza della legislazione* (*Science of legislation*), with the clearest references being those in the fourth volume, dedicated to criminal law (1781–1782), in which the ideas of Vattel emerge in relation to the questions of the proportionality between punishment and crime,[1] and of crimes against the state.[2] Filangieri introduced a particular category of crimes, namely those against society and the state made up of citizens, and his words on this matter were entirely in keeping with the ideas of Vattel: "Every civil society supposes the existence of a constitution and of a legal entity that represents the sovereignty. Whatever this constitution may be, whoever this representative of sovereignty is, every newborn citizen is charged with the duty to preserve unharmed the constitution of the government and to defend the one that represents sovereignty."[3]

Archival Abbreviations
ASM Archivio di Stato di Modena, Modena, Italy
ASV Archivio di Stato di Venezia, Venice, Italy
MCV Museo Correr, Venezia, Venice, Italy

© The Author(s) 2020
A. Trampus, *Emer de Vattel and the Politics of Good Government*,
https://doi.org/10.1007/978-3-030-48024-0_7

This language was taken up throughout the Italian peninsula thanks to the extraordinary dissemination of *Scienza della legislazione*, which revealed an evolution in the use of Vattel's work in line with the need to justify and legitimise the political changes made during the crisis of the *ancien régime*. Traces of Vattel could also be found in, for example, the Italian manuscript translation of an article entitled *Troubles de la République de Boulogne* published in 1782 in the *Journal des gens du monde* in Frankfurt. This concerned the opponents of the court of the pope, making reference to the "fundamental constitution of government" and the freedom of Bologna—free "as are the modern nations." Pietro Verri, in his capacity as a democrat and supporter of the Convention of Paris, was even more explicit when, in 1792, just after the Revolution, he wrote: "Vattel's theories seemed true to me thirty years ago and I now persevere the same opinion. No great change has ever occurred without a great shock and much disorder."[4]

FROM DUTCH TO VENETIAN NEUTRALITY

The demarcation between proper trade and the cynical abuse of the association of commerce with freedom was an important theme in Venice, where the Dutch republic was seen to represent the latter. Precisely this association found its way into the Venetian translation of the *Droit des gens*, which contained a false place of publication, Lyon, and was published in three volumes between 1781 and 1783.[5]

While Vattel had used the example of the Dutch when outlining his long-term theory of commercial integration between states, the Venetian reception of these chapters, twenty-five years later, involved a severe critique of Dutch trade policy. Such an assessment featured in Carlantonio Pilati's *Lettres sur la Hollande*. Pilati—a controversial figure due to his adherence to the Illuminati order—had an important role as a cultural mediator between the United Provinces and Italy—mainly Venice—and promoted the plan to publish a new translation of Vattel's text.[6] His belief that by the end of the eighteenth century the Dutch had been reduced to a nation of rent-seekers who profited from "Jealousies among Princes" echoed throughout the annotations of the Venetian edition.[7]

Chapter Seven of the *Droit des gens* frequently referred to Grotius in support of the obligation of neutral vessels to accept inspections of cargo suspected of containing contraband.[8] Whereas the Italian translation by Loschi was generally literal, its rendering of this chapter included numerous additional notes that were not clearly marked as such, and offer commentary rather than explanation.[9] The additions argued that the principles of international law put forward by Grotius and the actual behaviour of the United Provinces during the Eighty Years' War diverged.[10] The apparent message of these notes was that the Dutch had failed to perfect themselves in line with their own true interest and in the process also obstructed Vattel's vision of how to "make of modern Europe a kind of republic [...] linked together by the ties of common interest."[11]

The editor who inserted the critical notes about Dutch trade was Lodovico Antonio Loschi, who we encountered in the previous chapter and about whom it is appropriate to give more biographical detail here. Born in Modena on 4 June 1744, the son of a ducal archivist, Loschi studied law in Bologna and became a member of the group gathered around the marquis Alfonso Vincenzo Fontanelli, a diplomat, politician and president of the Biblioteca Estense and the University of Modena. Following a personal scandal, in around 1770 Loschi was forced to leave Modena and set up shop in Venice, where he soon involved himself in a vortex of activity revolving around frantic relationships within aristocratic, cultural and editorial circles. From 1771 he collaborated intensively with the printer Giovanni Vitto, the most famous fruit of which was the republication of Filangieri's *Scienza della legislazione*. Upon his return to Modena in 1788 Loschi became a professor of moral philosophy.

Curiously, the Venetian edition of the *Droit des gens* stood out among Loschi's commercial, journalistic and editorial activities because it was the result of a commission, as is derived from Loschi's correspondence. The initiative may have come from Andrea Tron, whose reputation Loschi had vindicated in the mid-1770s when Tron was accused of supporting Pilati and the Masonic lodges. The first known collaboration with Tron came in 1773 when Loschi assisted Gasparo Gozzi in the preparation of a poetic homage for the installation of Andrea Tron as Procurator of San Marco.[12] For the same occasion Loschi also commissioned the well-known adventurer Ange Goudar[13] to write a *Discours oratoire contenant l'éloge de son excellence monsieur le chevalier André Tron*. In 1780 Loschi was also requested by Tron to commission Goudar to counteract Dutch claims in the early stages of the famous Chomel-Jordan insurance controversy (discussed in the next chapter), in the direct aftermath of which Loschi embarked upon his translation of Vattel.

Loschi's connection to Tron is particularly significant, since the latter, together with Angelo Querini,[14] in the years before his death orchestrated a project to save the Venetian Republic of which the development of the famous *Codice per la veneta mercantile marina* was a key part.[15] Notably, this document changed Venice's policy of neutrality from the abstinence of 'exact neutrality' to a form of 'active neutrality,' and just before his death Tron delivered a public speech on the development of Venetian manufactures that was presumably part of the same overall plan. In the light of these initiatives the Vattel translation served as a legal manifesto of Venice's renewed self-understanding in the context of the interstate system, and provided intellectual support highlighting the discrepancy between politicised commercial rivalry (which the Dutch had promoted and subsequently become victims of) and the properly organised and long-term viable trade that Venice needed to develop.

The Venetian vision of reform appears to have adapted Vattel's idea that Britain might act as a protector of the position of the small states in the balance of power, changing it to suit the Italian context, with its patchwork of small states with limited autonomy. Venice itself, although nominally an independent trade republic until 1797, for most of the eighteenth century lived under the shadow of the Austrian Habsburg influence in Italy. As for Andrea Tron, from the 1740s he came to believe that Dutch politics were outdated and to see London, Paris and Vienna as the centres of modern monarchical power with which small states like Venice ought to develop relations. For Tron figures like Jacques Necker embodied the ideal of a model politician serving a modern monarchy who, based on an understanding of the distinction between proper and abusive trade could help stabilise the balance of power. Tron's ideas therefore appear to have resembled those of Bynkershoek' *Questions of Public Law* on this point.

The idea that the publication of the *Droit des gens* in Venice was not about the need to make a political choice between attempting to gain support from Britain, France or Austria, but was instead about waking up the Venetian patriciate to a reality that had not yet gained ground in the Republic, is confirmed by the involvement of the Anglophobic adventurer and political writer Ange Goudar.[16] In 1773 Goudar had, in the aforementioned text written for Tron, emphasised the difference between true ambassadors like the Tron family, with their expertise in international politics, and those who "se renfermant dans l'enceinte de la Patrie." Similarly, in 1775, writing in an attack on the reforms proposed by the institution of the Correttori della legge, Goudar described the inability of Venetian patricians to transform Venice into a functioning part of the balance of power.[17] Rather than complain about luxury and curb the excesses of the famous carnival, he argued that the real challenge was to restore Venetian commercial life. While the balance of power prescribed that each state should have a share of European political power as big as it could digest, Britain's appetite was bigger than its stomach, while Venice had starved for sixty years, a period during which it lost any right to compare itself to the ancient trading power of Carthage.

Also belonging to this period was the publication of *Il diritto delle genti, ovvero principii del diritto naturale applicati alla condotta e agli affari*, Loschi's translation of the *Droit des gens*.[18] Despite frequently being cited in studies of eighteenth-century Italian culture, Loschi has been left almost entirely in the shadows by twentieth-century historiography, which has mentioned him only in passing and credited him with no more than a secondary role in the intellectual panorama of his age.[19] Yet if one considers the legacy of certain documents preserved in the State Archives of Modena, which include seven large binders containing his surviving correspondence written over a fifty-year period,[20] one has the impression of a man of letters who was strongly committed to creating a role for himself in the Enlightenment debate: an indefatigable translator and journalist, then a university teacher, and ultimately a protagonist (albeit not of the highest order) in the political life of the three years of the Italian Republic and then the Napoleonic Kingdom of Italy.

In September 1764 Loschi was amongst the first people to read and disseminate the *Dei delitti e delle pene*, whose author was at that time still unknown but who Loschi immediately undertook to defend from the critics, praising both the content and the style of his work.[21] Loschi's discovery of Beccaria's work and his preoccupation with the contemporary legal and political debate in Europe thus began to be accompanied by a strong predilection for French Catholic and Jansenist culture, as is clear from his early translation of *Abregé de l'histoire ecclésiastique* by Bonaventure Racine and his interest in the school at Port Royal. These two passions, for the legal Enlightenment and Jansenist culture, were unified by a rationalist ethic that could support his ambitions, which were probably focused on an academic career. It was for this reason that he soon began to collaborate with the printer Giovanni Montanari, who specialised in the publication of legal works for universities: in 1768 Loschi edited the *Dissertazione intorno le ragioni di promulgare, o abrogare le leggi* by Frederick II of Prussia, which appeared at almost the same time as the Modenese reprint of *Delle virtù e de' premi* by Giacinto Dragonetti, a young student of Antonio Genovesi.[22] That same year also saw the publication, under the title of *Il figlio naturale, ossieno le pruove della virtù,* of his translation of Diderot's comedy *Le fils naturel, ou les épreuves de la vertu.*[23] In the following two years, having by then become Montanari's consultant on legal works, Loschi became a close friend of Agostino Paradisi and, closer still, to Lazzaro Spallanzani, a consultant to Montanari on scientific disciplines who not long after took up a post at the University of Pavia.

In 1770 Loschi's life took a turn for the worse: having been accused of helping a noblewoman escape from Carpi, the town in which she had been confined by order of the Duke, or perhaps having simply fallen victim to calumny, he was forced to leave Modenese territory and find refuge in Venice, where he remained in a sort of exile for more than fifteen years. There he initially lived in poverty and near secrecy, all the while lamenting his "great financial mess," his "obscurity" and "shame,"[24] until, in the mid-1770s, he began a career as a journalist, primarily working for gazettes and local printers as a translator from English and French.

By 1771 he was already working with the printer Giovanni Vitto thanks to the publication of the *Ragionamento intorno l'amministrazione della giustizia criminale pronunciato dal signor Servan,* a translation from French of a noted text of the *philosophes.*[25] In the mid-1770s he played a part in vindicating and defending the Enlightenment reputation of Tron, who at the time was accused, by Giuseppe Baretti amongst others, of supporting and protecting the freemason Carlantonio Pilati.[26] It was in this context, when the ecclesiastical reforms promoted by Tron seemed to symbolise a break between the culture of the nobility and jurisdictional and openly reformist politics, that Loschi began to strengthen his ties with Jansenist and reformist groups, publicly paying homage to Tron during his campaign for election to the post of procurator of San Marco in a poem entitled *Andreae Trono Equiti, dum Marcianae Basilicae Procurator electus in loci possessionem veniret, ospiti in Urbe Ludovici Antonii Losci Mutinensis gratulatio.*[27] For the same occasion he commissioned the celebrated adventurer Ange Goudar to write the *Discours oratoire contenant l'éloge de son excellence monsieur le chevalier André Tron cy devant ministre extraordinaire en Holande et ambassadeur aux cours de Versailles et de Vienne.*[28] In this work Goudar declared in the courts of Europe that he had been personally acquainted with Tron, seemingly in an attempt to remain in Venice by having Tron grant him some kind of trade appointment.[29] Loschi's correspondence reveals the background to this collaboration, recounting that he had procured Goudar work and money with the promise of receiving a percentage of his future earnings, a pledge that Goudar failed to keep.

In those years Loschi published with Vitto the *Saggio sopra il carattere i costumi e lo spirito delle donne ne' varii secoli, del sig. Thomas dell'Accademia francese. Traduzione italiana corredata di annotazioni storico-critiche; ed accresciuta di una Lettera dell'abate Conti P.V. intorno lo stesso argomento* followed by *Le lamentazioni, ossieno le Notti* by Young,[30] the Italian version of a celebrated text that recalled the percipience of Rousseau and suggested to the Venetian reader a parallel with Angelo Querini's Masonic home in Alticchiero, which was famous for its garden created in accordance with a kind of Masonic allegorical moral programme that brought to mind the contemplation of eternity and the regeneration of the virtuous man.[31] The *Lamentazioni* was followed by other translations such as *Il Trionfo della Religione sopra l'Amore, poemetto Inglese d'Odoardo Young* e *Il savio in solitudine o sieno religiose meditazioni sopra diversi soggetti di Edoardo Young.* As the seventies gave way to the eighties Loschi prepared the edition and annotation of *Dell'origine, de' progressi e dello stato attuale d'ogni letteratura* in twelve volumes by Juan Andrés and, as editor, himself published a collection of letters by Esprit Flechier, Bussy Rabutin and Madame de Sevigné entitled *Choix de lettres françoises à l'usage des collèges & de tous ceux qui apprennent la langue Françoise,*[32] which was extended and republished in Italian in 1808 with the title *Il segretario perfetto, ovvero modelli di lettere di vario argomento.* Finally, he was the author of a *Discorso sopra il sermone poetico* based on the *Sermoni* by Cosimo Mei.

However, the undertaking that kept him most occupied and of which—as can be seen from his letters—he was most proud was the translation in five volumes of the *Eléments d'histoire générale* by Millot, which appeared under the title *Elementi di storia generale antica e moderna opera scritta nell'idioma francese dal sig. Abate Millot; recata nell'italiano da Lodovico-Antonio Loschi, con varie aggiunte ed annotazioni* (1777–1781).[33] This went on to be republished in 1786, 1803 and 1816.[34] In the mid-1780s Loschi intensified his collaboration with the *Progressi dello spirito umano nelle scienze e nelle arti*, a journal with Jansenist sympathies that was often critical of the *Encyclopédie* project and of d'Alembert.[35] Its editor was Father Antonio Contin, and it represented the most advanced point of Venetian Jansenism, which, as Venturi observed, still continued to side with the "most independent men in Venetian culture, from Pilati to Fortis."[36]

Having returned to Modena by 1788, Loschi became professor of ethics and morals at the university: he was at that time the founder of a number of institutes of ethics.[37] In 1796, after the fall of the Estense government, he held posts on the education and public instruction boards of Modena and Bologna, then in Milan with the Cispadane and Cisalpine republics, up to the point of becoming, in 1797, a member of the Constitution Committee with Luigi Lambertenghi, Alfonso Longo and Francesco Melzi d'Eril, and he was a signatory to the Cisalpine constitution.[38] After the battle of Marengo he served as the provisional president of the Municipality of Modena, then afterwards retired into private life. He died on 18 August 1811.[39]

LEGAL KNOWLEDGE AND THE RENEWAL OF POLITICAL LANGUAGE

Loschi's main passion was always for the legal and political debates that surrounded the crisis of the *ancien régime* and the late Italian and European Enlightenment. As has already been mentioned, in his youth he had been inspired by the work of Beccaria and had collaborated in Modena with Montanari, the publisher of university legal texts. He had edited the *Dissertazione intorno le ragioni di promulgare, o abrogare le leggi* by Frederick II of Prussia around the same time as the reprinting of Dragonetti's *Delle virtù e de' premi*, thereby reflecting Genovesi's widespread fame. The latter's work was becoming an important vehicle for the transmission of new ethical and legal knowledge that would have a profound effect on the emerging generations of intellectuals and would draw Catholic culture and Enlightenment aspirations into dialogue.[40]

From the 1760s onwards Loschi therefore began to play a key role in the renewal of Italy's science of law, promoting the circulation and the dissemination of important Italian and foreign texts thanks to an extensive network of friends and correspondents and also his own rare ability to discern the public's tastes and concerns. An early demonstration of this was the translation of *Discours sur l'administration de la justice criminelle prononcé par M. S***, avocat général* by Joseph Michel Antoine Servan, published in 1771, but the definitive thrust into these projects occurred at the beginning of the 1780s when he suggested a project to the Venetian printer Rinaldo Benvenuti which included a new edition of *Dei delitti e delle pene* with observations by Voltaire, and various other books concerning the debate that ensued from Beccaria's essay.[41] Loschi's list of associates—which in the end ran to over fifty people from all over Italy—helped him to establish a chain of buyers and people interested in editorial initiatives in the fields of jurisprudence and politics.[42] Following his project's success he began work that same year on an Italian version, again with the presses of Benvenuti, of the *De tormentis* by Salvatore Venturini of Lucca, which had appeared anonymously in 1766 and was inspired largely by Beccaria. That work, however, never saw the light of day.[43] Undaunted, two years later Loschi dedicated himself to a new edition of the *Dei tributi*, also by Venturini, which claimed to be printed in Munich though in reality was from Venice.[44] There followed the translation of Vattel, to which we shall come later, as well as a new edition of the *Scienza della legislazione* by Gaetano Filangieri that he edited and which was published by Vitto.

The republication of the *Scienza della legislazione* was without doubt the enterprise that raised the profile of Loschi and Vitto to its highest level on the Italian and international editorial scene. Loschi had sensed immediately, from the first and halting Neapolitan edition, the interest that the *Scienza* would elicit in the public, and was fully aware of the limits and defects of the first two editions, which contained typographical errors and had too few copies.[45]

Following this, in 1787 he prepared a Venetian edition of the *Legge sulla riforma della Legislazione criminale in Toscana* and the Josephine Code,[46] and a year later, again for Vitto's printing-house, he oversaw the republication of the *Progetto di un nuovo codice giudiziario nelle cause civili* by Francesco Vigilio Barbacovi, edited a short time earlier in Trento.[47]

THE GENESIS OF LOSCHI'S TRANSLATION AND ITS EDITORIAL CONTEXT

The Venetian translation of Vattel's work bore the title *Il diritto / delle genti / ovvero / principii / della legge naturale / applicati alla condotta e agli affari / delle Nazioni e de' Sovrani / opera / scritta nell'idioma francese / dal Sig. di Vattel / e recata nell'italiano / da Lodovico Antonio Loschi,* and was published in three octavo volumes. It was not based, as we have often been led to believe, on the first edition of 1758, but rather on the 1773 expanded and revised edition published in Neuchâtel. Judging by the archival documents relating to the printing licences, it would seem that it was completed relatively quickly, between the summer of 1780 and the autumn of 1781.[48] The publication of the first manuscript was granted authorisation on 22 March 1781, with the false place-name of Lyon, and was conceded to the printer Giovanni Gatti, in accordance with the judgement of the proof-inspector Gasparo Gozzi. The second volume, later subdivided into three volumes, obtained authorisation on 22 September 1781, as recommended by the inspector Cosimo Mei. The speed of the publication was matched by that of the translation, which was such that in his letter to Mei of 22 October 1782, which accompanied the third volume, Loschi stated that he had had neither time nor opportunity to read it again or correct it.[49] The entire process had to be completed by the late spring of 1783, as shown in a letter from Loschi to Gatti of 8 June 1783, in which he referred to a sum received for payment of the translation,[50] and other letters to Appiano Buonafede of 21 June 1783, in which he announced that he would soon receive the first copies of the book,[51] and also in a letter from Buonafede to him, dated Rome, 13 December 1783, confirming that they had arrived.[52]

The *Droit des gens* was thus translated in haste, which implied that those involved were confident that authorisation to publish it would be given, even though prudence suggested that, with the possibility that it might not, it would be wise, though false, to state that the place of publication was Lyon, albeit "with public approval and privilege." Such haste can best be interpreted in the specific context of Venetian foreign policy, oriented towards a last attempt to reacquire certain positions within the European international system by transforming the Republic's age-old policy of 'perfect neutrality' into an active neutrality that could seize the opportunities arising from trade wars. The intellectuals close to Tron, who in the years just before had supported his reformist and jurisdictional policies, were not unaware of these efforts, and favoured the meeting between exponents of the Enlightenment world and those of reformist and Jansenist circles. It was not by chance that the idea of nominating Pilati, who was close to Tron at that time and often in Venice, to the post of *Consultore di Stato* was formed. Pilati was returning to Italy, bringing with him a store of experiences and literature linked to the Enlightenment interpretation of Grotius's natural law and of that developed in Switzerland. Moreover, thanks to his editorial ventures, he was often active in the book trade between Switzerland and Venice. It would therefore not be surprising if it were those circles, which Loschi frequented, that suggested the translation of the *Droit des gens*.[53]

Each one of the three volumes of the printed editions was dedicated to a young Venetian pupil of the College of Modena and member of the Bollani family. These were the brothers Giovanni, Jacopo and Antonio, sons of the late Girolamo.[54] As far as can be ascertained from the undated draft letter which he sent them with the first volume of the work, Loschi was in contact with at least one of the brothers and may have been their tutor and supported their studies in Modena.[55]

In order to fully understand the background against which the translation was carried out one must also consider a series of other factors that go beyond the conventional dedication to the Bollani brothers. First of all there were the requirements of the book trade in the 1780s, which was more and more involved with political and legal texts, particularly as the new political battles and the outcome of the Seven Years' War accelerated the crisis of the *ancien régime*. Then there was the free-for-all between publishers, which led to an uncompromising race for the publication of original texts or for their translation. And then there was the growing respect for the role of the translator as an active participant in the transformation of a literary work. Finally, one should not forget the rivalry among translators who had increased in number and who offered publishers texts of varying quality at competitive prices: amongst them were many former Jesuits, not a few from the Iberian peninsula, who dedicated themselves to this activity so as to make ends meet.[56]

From all of this one can understand the context of, and the events that ran alongside, the birth of *Il diritto delle genti* and the relations between Loschi, the printer Gatti and the intellectuals of the period. Further evidence can be found in the contemporaneous translation of the text of Burlamaqui, *Principes du droit naturel*, with which Gatti was occupied shortly before the publication of Vattel. The translation of Burlamaqui had been the work of Count Benedetto Crispi, with the annotation being made by the former Jesuit Luciano Galissà y Costa.[57] Crispi, from Ferrara, was one of the main patrons of ex-Jesuit exiles in that region, and was a friend and translator of Juan Andrés and well-known for versions of works by Esiod, Mably and Calonne.[58] In that instance the real objective of the translation of Burlamaqui was to acquire a work belonging to the tradition of pre-Enlightenment lay natural law and bend it to Catholic ideology, whereas, from Gatti's point of view, it was to provide an alternative to the eight-volume translation of Burlamaqui that had appeared at the same time in Siena based on the edition from the presses of Fortunato Bartolomeo de Felice at Yverdon in 1766–1768.

However, that the Italian public's chief concern was the quality of translations can be seen from certain letters that Loschi found himself writing on Gatti's behalf in defence of the version of Burlamaqui following criticisms levelled against him by readers.

Loschi's Strategies: The Annotation and Choices of Translation

Perhaps it was precisely those events surrounding the translation of Burlamaqui that made Loschi reticent about laying claim to the merits of his translation: the version of the *Droit des gens* included in fact only a very brief *Avvertimento del volgalizzatore* to explain that the translator had chosen to prioritise the purpose and didactic efficacy of the contents over elegance of style. He therefore suggested an approach to the work of Vattel that kept in mind the educative intent and the characteristics of the prose; furthermore, one should not forget that the language of the Swiss author could not meet the level of purity and propriety that one might expect from a native Frenchman, which obviously ended up conditioning the results of the translation.

In reality the work of the translator was not limited simply to producing an Italian version of the text; Loschi also provided explanatory notes, especially in the first volume. The notes were by no means few and related to certain key questions of the 1770s and 1780s debate and are worth noting. For instance, when translating Vattel's observation of the weakness of the Polish constitution, Loschi added a note observing how events following 1776 had confirmed this fragility.[59] More incisive was the position he took up in relation to Vattel's consideration on the undue influences of a foreign nation on the right of royal succession. The question referred to the power of the Catholic Church to deny the sacrament of marriage or to prevent the divorce of a foreign prince even if this were dictated by the reason of state. Vattel had explained that this power should be nullified since it amounted to interference in the internal affairs of an independent sovereign nation, but in response Loschi observed that Vattel's thoughts expressed the Protestant point of view and that the impediment on marriage derived not from the choices of the Catholic Church but from divine natural law.

There followed a long list of sources taken from Roman law and canon law that reached a conclusion reaffirming the authority of the Pope alone for the concession of the dispensation of diriment impediments to marriage.

However, the question was held to be particularly important and Loschi returned to it when commenting on another affirmation by Vattel, who had found in legislation passed by Ludwig of Bavaria an attempt to define as prejudicial to imperial authority any action of the Church aimed at limiting the Emperor's prerogatives, particularly on the matter of issuing and annulling marriages. Loschi once more based his views on a long series of canonical sources.[60]

It is clear that Loschi's first responses related essentially to the more markedly anti-curial positions of Vattel. This was a somewhat understandable attitude, given the prevailing orientation of the public and the potential intrusion of the censor. It is therefore unsurprising that similar observations by Loschi emerged as comments on those pages in which Vattel asserted the freedom of religion and of conscience to be natural and inviolable rights. Here Loschi observed that freedom of conscience should nevertheless not enter into conflict with the right and duty of the Catholic Church to provide for civil society in order that the right of free conscience was identified with natural law so that one had to keep to oneself any intimate thought that, if made public, could be harmful to society.[61]

The question of freedom reappeared in other points of the work and again Loschi was a scrupulous annotator, for example on the subject of personal freedom where Vattel reasoned on the duties that the people and the nation had as regards the obligation to agree on the construction and maintenance of public works, and the rights of governors to require the people to render personal service in support of the state's needs. Whereas Vattel believed this practice to be useful and necessary, Loschi maintained that sovereigns should refrain from promoting public works aimed only at satisfying their personal vanity.[62] Once more, the theme of freedom emerged in relation to the right of trade; in the second volume of the work Vattel had in fact spent time considering this issue and the right of a nation to choose whether to trade with other nations. He pointed out, as a clear example of the violation of this principle, the behaviour of Spanish settlers towards Native Americans—who were free people—which had been driven by avarice and commercial exploitation. To this reflection Loschi added a note defending the Spanish conquests and asserting that the crimes and misdeeds of a few could not be attributed to the will of sovereigns and governors.[63]

Finally, Loschi inserted another note on the issue of the responsibility of nations for acts committed by their members. Vattel had in fact confronted the problem of the war of privateers as an obstacle to the freedom of trade, especially in the Mediterranean, and had theorised the right of Christian nations to ally themselves against Barbary states when the latter made no attempt to eliminate the phenomenon. Loschi, however, turned the argument around, suggesting that the privateer war could be seen as a means of subsistence for a nation and, as such, should be considered an expression of that nation's freedom and independence.[64]

One can see, in connection with this, how the different attitudes towards the 'Barbary' nations in Vattel and in Loschi substantially reflected the different position held by France and the Italian states respectively in relation to the very existence of the Barbary regencies, even if this argument was not developed by the two authors. The scant other notes were merely explanatory and, as such, did not involve significant interventions or formal modifications. It thus seems apparent that on the whole the translator's interventions through his annotations of a text spread over three volumes were fairly limited and always clearly recognisable thanks to the reference made by Loschi to the *Nota del traduttore*. Furthermore, it can also be observed how there were more interventions in the first volume compared to the other two, the former being that dedicated not to the study of international law, but to the examination of the constitutional nature of states and governments. This was perhaps no coincidence, given that the first volume was the one published before authorisation had arrived from the inspectors at the Padua office, from which fact one might deduce that the larger number of annotations might, on the one hand, have served to preclude hostility from the mostly conservative Venetian public, and, on the other, might have had the precautionary (and thus self-censoring) aim of avoiding the expected revision of the censors.

As far as the translation strategy is concerned, however, the choices taken by Loschi are not particularly surprising. Generally speaking, his work adhered to his initial assurance of producing a translation that was as literal and true to the original as possible. This can be confirmed by looking at the construction of sentences, as well as the frequent use of emphasis, particularly where the Italian political lexicon did not yet offer adequate terms to translate those used by Vattel. Moreover, there is a notably increased caution within the translation when rendering in Italian the paragraphs relating to the rights of neutrality and the republican nature of states, an argument that was particularly sensitive for the Venetian culture, and thus also in that of Vattel's definition of Europe and its system of balance, which conformed to a formula suggested by Christian Wolff and repeated by Voltaire in *Le siècle de Louis XIV*, according to which the continent was "une espèce de République, don't les membres indépendans, mais lies par l'intérêt commun". Loschi translated this as "una specie di Repubblica, i cui membri indipendenti, non legati dall'interesse comune, riunisconsi per mantenervi l'ordine."[65] Loschi altered the "mais" used adversatively by Vattel, in order to underline how the independence of European nations fell away before the common interest, to a negation ("non"), which ended up reinforcing the contrary idea that the interest of each individual state came before the common one.

However, even in this case the translator's interventions were generally recognisable and did not alter the original text. This allows us to understand why the circulation in Italy of this translation of Vattel's work, which was originally linked to a Protestant and rationalistic context, did not encounter any particular obstacle, as even the diffusion of the work in its original language demonstrated. The *Diritto delle genti e della natura* therefore appeared as a text open to many levels of reading, useful both to the Enlightenment culture and to more conservative, Catholic and Jansenist circles. However, the value of the translation was linked to the adaptation of Vattel's lexicon to Italian circumstances, into which the Swiss author succeeded in introducing a markedly modern language, including new meanings to words and phrases such as 'nation,' 'homeland,' 'the people,' 'the constitution of state,' 'fundamental law,' and 'legislative authority.'

NOTES

1. Gaetano Filangieri, *La scienza della legislazione. Edizione critica*, vol. 4, *Delle leggi criminali. De' delitti e delle pene*, ed. Gerardo Tocchini and Antonio Trampus (Venice: Edizioni della Laguna 2004), 138.
2. Filangieri, *La scienza della legislazione*, 43 and 159.
3. Filangieri, *La scienza della legislazione*, 159.
4. Carlo Capra, *I progressi della ragione. Vita di Pietro Verri* (Bologna: il Mulino 2002), 562–563.
5. *Il diritto delle genti*, ed. Loschi.
6. See the biography in Giuseppe Pilati, *Cenni su la vita e su le opere di Carlo Antonio Pilati stesi per la prima volta coll'aiuto di documenti da un Trentino* (Rovereto: Sottochiesa, 1874); Maria Rigatti, *Un illuminista trentino del*

secolo XVIII: Carlo Antonio Pilati (Florence: Vallecchi, 1923); Franco Venturi, *Illuministi italiani* 3, *Riformatori lombardi, piemontesi e toscani* (Milan-Naples: Ricciardi, 1958), 620–40; Venturi, *Settecento riformatore* 2 (Turin: Einaudi, 1976), 250–325; *Settecento riformatore* 4/2 (Turin: Einaudi, 1984), 543–63; Serenella Armellini, *Libertà e legislazione. Il riformismo di Carlantonio Pilati* (Milan: Jaca Book, 1991); Renato Gaeta, *Carlo Antonio Pilati. Dalle esperienze culturali europee al riformismo trentino (1760–1802)* (Venice: Deputazione di Storia Patria per le Venezie, 1995); Stefano Ferrari and Gian Paolo Romagnani, eds., *Carlantonio Pilati: Un intellettuale trentino nell'Europa dei lumi* (Milan: FrancoAngeli, 2005).

7. Venturi, *Settecento riformatore*, 4/2, 525–543; Koen Stapelbroek and Antonio Trampus, "*Commercial reform against the tide: Reapproaching the eighteenth-century decline of the republics of Venice and the United Provinces*" in *History of European Ideas* 36 (2010): 192–202.

8. See Vattel, *Droit des gens*, book II, ch. VII.

9. See Vattel, *Il diritto delle genti*, ed. Loschi, vol. 3, 82–109.

10. See Vattel, *Il diritto delle Genti*, ed. Loschi, vol. 3, 92–3.

11. Vattel, *Droit des gens*, book III, ch. III, § 47.

12. *Andreae Trono Equiti, dum Marcianae Basilicae Procurator electus in loci possessionem veniret, ospiti in Urbe Ludovici Antonii Losci Mutinensis gratulatio*, manuscript in ASM, Carte Loschi, 37/1 nr. 35.

13. On Goudar see Jean-Claude Hauc, *Ange Goudar: Un aventurier des Lumières* (Paris: Honoré Champion, 2004).

14. *Le lamentazioni, ossieno le Notti d'O. Y. coll'aggiunta di altre sue operette, libera traduzione di Lodovico Antonio Loschi* (Venice: Vitto, 1774), 77, the Italian version of a celebrated text that recalled the percipience of Rousseau and suggested to the Venetian reader a parallel with Angelo Querini's Masonic lodge in Alticchiero near Padua.

15. Maria Borgherini-Scarabellin, *Il Magistrato dei Cinque Savi alla Mercanzia dalla istituzione alla caduta della Repubblica: studio storico su documenti d'archivio* (Venice: Deputazione di storia patria per le Venezie, 1926). For instance, ASV, V Savi alla Mercanzia, busta 39, April 1727, Codice per la veneta mercantile marina, Titolo X 'Consoli,' art. 10, f.121–2). For the genesis of the *Codice* see Giorgio Zordan, *Il Codice per la Veneta mercantile marina* (Padua: Cedam, 1987).

16. Goudar was also involved in a scheme designed by Forbonnais to reform international trade competition between France and Britain and the role of the smaller and neutral states; see Antonella Alimento, "From Privilege to Equality: Commercial Treaties and the French Solutions to International Competition (1736–1770)," in *The Politics of Commercial Treaties in the*

Eighteenth Century. Balance of Power, Balance of Trade, eds. Antonella Alimento and Koen Stapelbroek (Cham: Palgrave Macmillan, 2017), 255.

17. Ange Goudar, *Plan de reforme, propose aux cinq correcteurs de Venise actuellement en charge. Avec un sermon evangelique pour elever la Republique dans la crainte de Dieu* (Amsterdam: n.p., 1775). See Francis L. Mars, "Ange Goudar, cet inconnu (1708–1791)," *Casanova Gleanings* IX (1966), 43–44.

18. Vattel, *Il diritto delle genti*, ed. Loschi.

19. Antonio Trampus, "Il ruolo del traduttore nel tardo Illuminismo: Lodovico Antonio Loschi e la traduzione italiana del *Droit des gens*," in *Il linguaggio del tardo Illuminismo*, ed. Antonio Trampus (Rome: Edizioni di Storia e Letteratura, 2009, 88–89.

20. Part of Loschi's correspondance was published by Giovanni Zannoni, *Una lettera inedita di Carlo Innocenzo frugoni a Lodovico Antonio Loschi* (Rome: Tip. Elzeviriana, 1895); Luigi Pucci, *Lodovico Ricci. Dall'arte del buon governo alla finanza moderna 1742–1799* (Milan: Giuffrè 1971), 207, 211–212.

21. William Spaggiari, *L'armonico tremore. Cultura settentrionale dall'Arcadia all'età napoleonica* (Milan: FrancoAngeli, 1990), 35, 48

22. Spaggiari, *L'armonico tremore*, 38; the reference to the work of Dragonetti is to *Delle virtù e dei premi* (Modena: Nella stamperia Montanari, 1768), VIII, 102.

23. Franco Venturi, *Settecento riformatore*, vol. 5/2, *L'Italia dei Lumi (1764–1790). La Repubblica di Venezia (1761–1797)* (Turin: Einaudi, 1990), 163.

24. Luigi C[agnoli], "Lodovico Antonio Loschi," in *Notizie biografiche in continuazione della Biblioteca modenese del cavaliere abate Girolamo Tiraboschi*, vol. 5 (Reggio: Torregiani, 1837), 335.

25. *Ragionamento intorno l'amministrazione della giustizia criminale pronunziato dal signor Servan avvocato generale al parlamento di Grenoble e recato dall'idioma francese all'italiano da L. A. Loschi* (Venice: Vitto, 1771). The manuscript of this translation, quoted by Spaggiari, *L'armonico tremore*, 51, is in ASM, Archivio per materie. Letterati-L. A. Loschi, 31/2.

26. C[agnoli], "Lodovico Antonio Loschi," 338; Giovanni Tabacco, *Andrea Tron e la crisi dell'aristocrazia senatoria a Venezia* (Udine: Del Bianco, 1983), 39.

27. ASM, Archivio per materie. Letterati-Lodovico Antonio Loschi, busta 31/7, nr.35, pages not numbered.

28. Ange Goudar, *Discours oratoire contenant l'éloge de son excellence monsieur le chevalier André Tron cy devant ministre extraordinaire en Holande et ambassadeur aux cours de Versailles et de Vienne* (Venice: Palese, 1773). On Goudar see Mars, "Ange Goudar, cet inconnu," 1–65; Gianfranco

Dioguardi, *Ange Goudar contro l'Ancien Régime* (Palermo: Sellerio, 1988); Hauc, *Ange Goudar*, 180–192.

29. See the letters of Andrea Tron to Goudar (4 January 1774 and 27 August 1774) in MCV, ms. Donà dalle Rose 462, fasc. VI, copialettere di Andrea Tron, quoted by Tabacco, *Andrea Tron*, 39–41.

30. *Le lamentazioni, ossieno le Notti d'O. Y.*

31. Eva Faber, "Die Ehe der Gräfin Giustiniana Rosenberg-Wynne (1737–1791)," in *Adel im "langen" 18. Jahrhundert*, ed. G. Haug-Moritz *et al.* (Vienna: Verlag Österreichische Akademie der Wissenschaften, 2009), 289–310.

32. Madame de Sevigné, *Choix de lettres françoises à l'usage des collèges & de tous ceux qui apprennent la langue Françoise* (Venice: Benvenuti, 1782); second edition Venice: Pasquali, 1788.

33. To Loschi was also attributed, incorrectly, the Italian version of Marc Antoine Laugier, *Histoire de la République de Venise* (Venice: Carlo Palese and Gasparo Storti, 1767–1769), which was in fact translated by Zaccaria Sceriman; see Venturi, *Settecento Riformatore*, V/1, 163.

34. Also published in Venice, respectively by the printers Storti, Giacomo Costantini and Molinari.

35. Franco Trentafonte, *Giurisdizionalismo, Illuminismo e massoneria nel tramonto della Repubblica veneta* (Venice: Deputazione editrice, 1984), 116–117. On Venetian journalism see Ilaria Crotti and Ricciarda Ricorda, eds., *Gasparo Gozzi. Il lavoro di un intellettuale nel Settecento veneziano* (Padua: Editrice Antenore, 1989) with essays by Piero Del Negro, "Gasparo Gozzi e la politica veneziana," 45–65 and Michele Cataudella, "Antilluminismo e progresso nell'ultimo Gozzi," 445–454. On Loschi as a journalist see also Carlo Capra, *Giovanni Ristori da illuminista a funzionario (1755–1830)* (Florence: La Nuova Italia, 1968), 70, 105.

36. Venturi, *Settecento riformatore*, vol. V/2, 224.

37. C[agnoli], "Lodovico Antonio Loschi," 339.

38. C[agnoli], "Lodovico Antonio Loschi," 340.

39. C[agnoli], "Lodovico Antonio Loschi," 333–341.

40. See Niccolò Guasti, "Un caso editoriale: la Diceosina di Antonio Genovesi" in Antonio Genovesi, *Della diceosina, o sia della filosofia del giusto e dell'onesto*, ed., Niccolò Guasti (Venice: Centro di Studi sull'Illuminismo europeo-Edizioni della Laguna, 2008).

41. See the edition Cesare Beccaria, *Dei delitti e delle pene. Novissima edizione di nuovo corretta, ed accresciuta coi commenti del Voltaire, confutazioni ed altri opuscoli interessanti di varj autori*, 2 vols. (Venice: Rinaldo Benvenuti, 1781).

42. The list of the associates was published in Beccaria, *Dei delitti e delle pene* (1781 edition), vol. 2, 227–246 and was studied by Gianfranco Torcellan, "Cesare Beccaria e Venezia," *Rivista Storica Italiana* LXXVI (1964), 747.

43. Spaggiari, *L'armonico tremore*, 52.

44. ASM, Archivio per materie, Letterati-L.A. Loschi, busta 31/6, letter from Venturini to Loschi of 27 November 1781, pages not numbered; see also Spaggiari, *L'armonico tremore*, 49. On Venturini see Franco Venturi, "Ritratto di Agostino Paradisi," *Rivista Storica Italiana* LXXIV (1962), 735.

45. Antonio Trampus, "La genesi e la circolazione della "Scienza della legislazione". Saggio bibliografico," *Rivista Storica Italiana* CXVII (2005), 309–359.

46. Spaggiari, *L'armonico tremore*, 40, 52

47. Maria Rosa Di Simone, *Legislazione e riforme nel Trentino del Settecento: Francesco Vigilio Barbacovi tra assolutismo e Illuminismo* (Bologna: il Mulino, 1992).

48. Patrizia Bravetti and Orfea Granzotto, *False date. Repertorio delle licenze di stampa veneziane con falso luogo di edizione (1740–1797)* (Florence: University Press, 2008), 237.

49. ASM, Archivio per materie. Letterati-L.A. Loschi, b. 31/1, nr. 24, pages not numbered.

50. ASM, Archivio per materie. Letterati-L.A. Loschi, b. 31/2, nr. 18, pages not numbered.

51. ASM, Archivio per materie. Letterati-L.A. Loschi, b. 31/3, pages not numbered.

52. ASM, Archivio per materie. Letterati-L.A. Loschi, b. 31/3, pages not numbered.

53. Franco Venturi, *Venezia nel secondo Settecento* (Turin: Tirrenia Stampatori, 1980), 128–129; Ferrari and Gian Romagnani, eds., *Carlantonio Pilati: un intellettuale trentino nell'Europa dei Lumi*, containing Antonio Trampus, "Dal giusnaturalismo alla politica del diritto: Carlantonio Pilati e l'Olanda," 158–192; Trampus and Stapelbroek, "Commercial reform against the tide," 92–202.

54. Giovanni Antonio Bollani, born on 13 June 1755, married Pasqua Domenica Fanton in 1796, see Francesco Schroeder, *Repertorio biografico delle famiglie confermate nobili e dei titolati* (Venice: Alvisopoli, 1830), 132–133.

55. ASM, Archivio per materie. Letterati-L.A. Loschi, b. 31/2, pages not numbered: "I am sending your. Excellency a V.E. a bundle with three copies of the first volume of the *Diritto delle genti* by Mr. Vattel vulgarised by myself; one for your excellency and the other two for your excellence's brothers."

56. Antonella Cancellier and Giuseppe Grilli, "La riflessione linguistica e traduttologica dei gesuiti in Italia: l'esempio di Masdeu," in *La presenza dei gesuiti iberici espulsi. Aspetti religiosi, politici, culturali*, eds. Ugo Baldini and Gian Paolo Brizzi (Bologna: Clueb, 2010), 577–586; Domenico Proietti, "La frammentazione dialettale e la situazione linguistico-culturale italiana nell'opera di Lorenzo Hervas y Panduro," in Baldini and Brizzi, *La presenza dei gesuiti iberici espulsi*, 587–608.

57. Lorenzo Hervás y Panduro, *Biblioteca jesuitico-española (1759–1793)*, ed. Antonio Astorgano Arajo (Madrid: Libris, 2007), 237.

58. Niccolò Guasti, *L'esilio italiano dei gesuiti spagnoli. Identità, controllo sociale e pratiche culturali* (Rome: Edizioni di Storia e Letteratura, 2006), 203–204, 281–282. See also Roberto Zapperi, "Edmund Burke in Italia," *Cahier Vilfredo Pareto 7–8* (1965), 38–40 and Franco Venturi, "Economisti e riformatori spagnoli e italiani del '700," *Rivista storica italiana* LXXIV (1962), 551.

59. Vattel, *Il diritto delle genti*, ed. Loschi, vol. 1, 33; compare with Vattel, *Droit des gens*, book I, ch. I, § 24.

60. Vattel, *Il diritto delle genti*, ed. Loschi, vol. 1, 77; compare with Vattel, *Droit des gens*, book I, ch. VI, § 61.

61. Vattel, *Il diritto delle genti*, ed. Loschi, vol. 1, 129–130; compare with Vattel, *Droit des gens*, book I, ch. XII, § 129.

62. Vattel, *Il diritto delle genti*, ed. Loschi, vol. 1, 106; compare with Vattel, *Droit des gens*, book 1, ch. IX, § 102.

63. Vattel, *Il diritto delle genti*, ed. Loschi, vol. 2, 25; compare with Vattel, *Droit des gens*, book II, ch. II, § 25.

64. Vattel, *Il diritto delle genti*, ed. Loschi, vol 2, 63; compare with Vattel, *Droit des gens*, book II, ch. V, § 78.

65. Vattel, *Il diritto delle genti*, ed. Loschi, vol. 3, 42–43.

BIBLIOGRAPHY

Most of the quotations of Vattel's *Droit des gens* come from the widely available English-language edition of 2008, which includes an introduction by Béla Kapossy and Richard Whatmore (Indianapolis: Liberty Fund) and maintains the text and English title *The Law of Nations: Or, Principles of the Law of Nature Applied to the Conduct and Affairs of Nations and Sovereigns* of the 1797 London standard edition.

Contemporary translations of primary sources are listed under the names of their authors. Full manuscript sources are referenced only in the notes for reasons of space.

ARCHIVAL ABBREVIATIONS

ASM   Archivio di Stato di Modena, Modena, Italy
ASV   Archivio di Stato di Venezia, Venice, Italy
MCV   Museo Correr, Venezia, Venice, Italy

PRIMARY SOURCES

PRINTED BOOKS

Beccaria, Cesare, *Dei delitti e delle pene. Novissima edizione di nuovo corretta, ed accresciuta coi commenti del Voltaire, confutazioni ed altri opuscoli interessanti di varj autori*, 2 vols. (Venice: Rinaldo Benvenuti, 1781)

Dragonetti, Giacinto, *Delle virtù e dei premi* (Modena: Montanari, 1768)

Filangieri, Gaetano, *La scienza della legislazione. Edizione critica*, 7 vols., dir. Vincenzo Ferrone, eds. Antonio Trampus *et al.* (Venice: Edizioni della Laguna 2004)

Genovesi, Antonio, *Della diceosina, o sia della filosofia del giusto e dell'onesto*, ed. Niccolò Guasti (Venice: Centro di Studi sull'Illuminismo europeo-Edizioni della Laguna, 2008)

Goudar, Ange, *Discours oratoire contenant l'éloge de son excellence monsieur le chevalier André Tron cy devant ministre extraordinaire en Holande et ambassadeur aux cours de Versailles et de Vienne* (Venice: Palese, 1773)

Goudar, Ange, *Plan de reforme, propose aux cinq correcteurs de Venise actuellement en charge. avec un sermon evangelique pour elever la Republique dans la crainte de Dieu* (Amsterdam: n.p., 1775)

Hervás y Panduro, Lorenzo, *Biblioteca jesuitico-española (1759–1793)*, ed. Antonio Astorgano Arajo (Madrid: Libris, 2007)

Laugier, Marc Antoine, *Histoire de la République de Venise* (Venice: Carlo Palese and Gasparo Storti, 1767–1769)

Sevigné, Madame de, *Choix de lettres françoises à l'usage des collèges & de tous ceux qui apprennent la langue Françoise* (Venice: Benvenuti, 1782; second edition Venice: Pasquali, 1788)

Vattel, Emer de, *Il diritto della natura e delle genti ovvero principii della legge naturale applicati alla condotta e agli affari delle nazioni e de' sovrani scritta nell'idioma francese dal Sig. di Vattel e recata nell'italiano da Lodovico Antonio Loschi* (Lyon [Venice]: n.p. [Vitto], 1781)

Vattel, Emer de, *Law of Nations: Or, Principles of the Law of Nature, Applied to The Conduct and Affairs of Nations and Sovereigns*, eds. Béla Kapossy and Richard Whatmore (Indianapolis: Liberty Fund 2008)

Young, Edward, *Le lamentazioni, ossieno le Notti d'O. Y. coll'aggiunta di altre sue operette, libera traduzione di Lodovico Antonio Loschi* (Venice: Vitto, 1774)

SECONDARY SOURCES

Alimento, Antonella, "From Privilege to Equality: Commercial Treaties and the French Solutions to International Competition (1736–1770)," in *The Politics of Commercial Treaties in the Eighteenth Century. Balance of Power, Balance of Trade*, ed. Antonella Alimento and Koen Stapelbroek (Cham: Palgrave Macmillan, 2017), 243–266

Armellini, Serenella, *Libertà e legislazione. Il riformismo di Carlantonio Pilati* (Milan: Jaca Book, 1991)

Borgherini-Scarabellin, Maria, *Il Magistrato dei Cinque Savi alla Mercanzia dalla istituzione alla caduta della Repubblica: studio storico su documenti d'archivio* (Venice: Deputazione di storia patria per le Venezie, 1926)

Bravetti, Patrizia and Granzotto, Orfea, *False date. Repertorio delle licenze di stampa veneziane con falso luogo di edizione (1740–1797)* (Florence: University Press, 2008)

Buckle, H. T., *History of Civilization in England*, 2 vols. (London: Longmans Green and Co., 1873)

Cagnoli, Luigi, *Lodovico Antonio Loschi*, in *Notizie biografiche in continuazione della Biblioteca modenese del cavaliere abate Girolamo Tiraboschi*, vol. 5 (Reggio: Torregiani, 1837), 333–341

Cancellier, Antonella and Grilli, Giuseppe, "La riflessione linguistica e traduttologica dei gesuiti in Italia: l'esempio di Masdeu," in *La presenza dei gesuiti iberici espulsi. Aspetti religiosi, politici, culturali*, eds. Ugo Baldini and Gian Paolo Brizzi (Bologna: Clueb, 2010), 577–586

Capra, Carlo, *Giovanni Ristori da illuminista a funzionario (1755–1830)* (Florence: La Nuova Italia, 1968)

Capra, Carlo, *I progressi della ragione. Vita di Pietro Verri* (Bologna: il Mulino 2002)

Crotti, Ilaria and Ricorda, Ricciarda, eds., *Gasparo Gozzi. Il lavoro di un intellettuale nel Settecento veneziano* (Padua: Editrice Antenore, 1989)

Di Simone, Maria Rosa, *Legislazione e riforme nel Trentino del Settecento: Francesco Vigilio Barbacovi tra assolutismo e Illuminismo* (Bologna: il Mulino, 1992)

Dioguardi, Gianfranco, *Ange Goudar contro l'Ancien Régime* (Palermo: Sellerio, 1988)

Faber, Eva, "Die Ehe der Gräfin Giustiniana Rosenberg-Wynne (1737–1791)," in *Adel im "langen" 18. Jahrhundert*, eds. G. Haug-Moritz et al. (Vienna: Verlag Österreichische Akademie der Wissenschaften, 2009), 289–310.

Ferrari, Stefano, and Romagnani, Gian Paolo, eds., *Carlantonio Pilati: Un intellettuale trentino nell'Europa dei lumi* (Milan: FrancoAngeli, 2005)

Gaeta, Rinaldo, *Carlo Antonio Pilati. Dalle esperienze culturali europee al riformismo trentino (1760–1802)* (Venice: Deputazione di Storia Patria per le Venezie, 1995)

Guasti, Niccolò, *L'esilio italiano dei gesuiti spagnoli. Identità, controllo sociale e pratiche culturali (1767–1798)* (Rome: Edizioni di Storia e Letteratura, 2006)

Guasti, Niccolò, "Un caso editoriale: la Diceosina di Antonio Genovesi," in Antonio Genovesi, *Della diceosina, o sia della filosofia del giusto e dell'onesto*, ed., Niccolò Guasti (Venice: Centro di Studi sull'Illuminismo europeo-Edizioni della Laguna, 2008), xi–lxxi

Hauc, Jean-Claude, *Ange Goudar: Un aventurier des Lumières* (Paris: Honoré Champion, 2004)

Mars, Francis L., "Ange Goudar, cet inconnu (1708–1791)," *Casanova Gleanings* IX (1966), 43–44

Pilati, Giuseppe, *Cenni su la vita e su le opere di Carlo Antonio Pilati stesi per la prima volta coll'aiuto di documenti da un Trentino* (Rovereto: Sottochiesa, 1874)

Proietti, Domenico, "La frammentazione dialettale e la situazione linguistico-culturale italiana nell'opera di Lorenzo Hervas y Panduro," in *La presenza dei gesuiti iberici espulsi. Aspetti religiosi, politici, culturali*, eds. Ugo Baldini and Gian Paolo Brizzi (Bologna: Clueb, 2010), 587–608

Pucci, Luigi, *Lodovico Ricci. Dall'arte del buon governo alla finanza moderna 1742–1799* (Milan: Giuffrè 1971)

Rigatti, Maria, *Un illuminista trentino del secolo XVIII: Carlo Antonio Pilati* (Florence: Vallecchi, 1923)

Schroeder, Francesco, *Repertorio biografico delle famiglie confermate nobili e dei titolati* (Venice: Alvisopoli, 1830)

Spaggiari, Wlliam, *L'armonico tremore. Cultura settentrionale dall'Arcadia all'età napoleonica* (Milan: FrancoAngeli, 1990)

Stapelbroek, Koen, and Trampus, Antonio, "*Commercial reform against the tide: Reapproaching the eighteenth-century decline of the republics of Venice and the United Provinces*" in *History of European Ideas*, vol. 36 (2010), 192–202

Tabacco, Giovani, *Andrea Tron e la crisi dell'aristocrazia senatoria a Venezia* (Udine: Del Bianco, 1983)

Torcellan, Gianfranco, "Cesare Beccaria e Venezia," *Rivista Storica Italiana* LXXVI (1964), 720–748

Trampus, Antonio, "Il ruolo del traduttore nel tardo Illuminismo: Lodovico Antonio Loschi e la traduzione italiana del *Droit des gens*," in *Il linguaggio del tardo Illuminismo*, ed. Antonio Trampus (Rome: Edizioni di Storia e Letteratura, 2009), 81–109

Trampus, Antonio, "La genesi e la circolazione della 'Scienza della legislazione'. Saggio bibliografico," *Rivista Storica Italiana* CXVII (2005), 309–359

Trentafonte, Franco, *Giurisdizionalismo, Illuminismo e massoneria nel tramonto della Repubblica veneta* (Venice: Deputazione editrice, 1984)

Venturi, Franco, "Economisti e riformatori spagnoli e italiani del '700," *Rivista storica italiana* 74 (1962a), 532–561

Venturi, Franco, "Ritratto di Agostino Paradisi," *Rivista Storica Italiana* 74 (1962b), 717–738

Venturi, Franco, *Illuministi italiani 3, Riformatori lombardi, piemontesi e toscani* (Milan-Naples: Ricciardi, 1958)

Venturi, Franco, *Settecento riformatore* vol. 2, *La Chiesa e la Repubblica dentro i loro limiti (1758–1785)* (Turin: Einaudi, 1976),

Venturi, Franco, *Settecento riformatore*, vol. 5/2, *L'Italia dei Lumi (1764–1790). La Repubblica di Venezia (1761–1797)* (Turin: Einaudi, 1990)

Venturi, Franco, *Settecento riformatore* vol. 4/2 *La caduta dell'Antico Regime 1776–1789* (Turin: Einaudi, 1984)

Venturi, Franco, *Venezia nel secondo Settecento* (Turin: Tirrenia Stampatori, 1980)

Zannoni, Giovanni, *Una lettera inedita di Carlo Innocenzo frugoni a Lodovico Antonio Loschi* (Rome: Tip. Elzeviriana, 1895)

Zapperi, Roberto, "Edmund Burke in Italia," *Cahier Vilfredo Pareto* 7–8 (1965), 38–40

Zordan, Giorgio, *Il Codice per la Veneta mercantile marina* (Padua: Cedam, 1987)

CHAPTER 8

Ships and Diamonds: Vattel Between Linguet and Casanova

The proceedings of the Berlin Academy of Sciences for the years 1788 and 1789, which were published with the date 1793, contain a discussion of Vattel's work recorded by Jean Pierre de Chambrier, a native of Neuchâtel and minister plenipotentiary in Turin of the King of Prussia.[1] The essay, which was reprinted a few years later as a standalone document by the famous typographer Bodoni in Parma, is interesting because it is one of the first reflections written by a diplomat, which is to say a practical executor of international policy, on the possibility of applying Vattel's theories and on the viability of his conceptual and interpretative tools.

We will return to this essay later, but first I want to anticipate a point which Chambrier cogitated on, namely the international consequences of the principle of national sovereignty and its associated rights. One particular problem considered by Chambrier concerned how a state should react when attacked not on its own territory, but through a threat to its citizens when abroad. If an unjust sentence were pronounced by a foreign court against a member of a state that might compromise the interests of the state itself, would that sentence constitute a threat to that state's

---

Archival Abbreviations

ASV   Archivio di Stato di Venezia, Venice, Italy

© The Author(s) 2020
A. Trampus, *Emer de Vattel and the Politics of Good Government*,
https://doi.org/10.1007/978-3-030-48024-0_8

sovereignty? And could the sentence be challenged without it signifying an attack on the sovereignty of the country that had imposed it?[2]

Chambrier had exposed a limit and a contradiction of Vattel's book, which in one part said that a sentence handed down by a state against the members of another state who were in its territory could not be questioned, but in another said instead that a prince had the right to intervene against an unjust judgement passed on his subjects by a foreign country. The case which Chambrier had in mind and which revealed the possible internal contradictions of the *Droit des gens* was that of the famous international dispute that a few years earlier had stirred up the European public: the Chomel and Jordan affair between the Dutch republic of the United Provinces and the Republic of Venice, and the international *affaire* of ships and diamonds that ensued from it.

LIBERTY AND THE BALANCE OF TRADE

Chambrier referred to this case because he had been personally involved in it when working as a diplomat in 1782.[3] It related to one of the most widely debated international questions of the late eighteenth century. Another participant was Giacomo Casanova, who was an observer watching the commercial interests of nations, and a journalist engaged by the ambassador of Venice to support the interests of the Republic in European public opinion when diplomatic means proved insufficient to resolve the dispute. Once again, lurking in the background was the problem of understanding what the balance of power actually meant at a time when the greatness of a nation was measured not so much by its domination of the seas as by its commercial policies.[4]

Casanova had begun to take an interest in international politics and the trade policy of the states when, soon after the Seven Years' War, he collaborated with Ange Goudar in the writing of *The Chinese Spy* (*L'Espion chinois*, 1764), a kind of epistolary novel which, in an ironic form and imitating the literary artifice of Montesquieu's *Persian Letters*, gave an account of a series of false dispatches through which a Chinese spy in London reported on the state of European commerce and on the reasons for British supremacy. In the imaginary letters, the protagonist, a Chinese mandarin, brought up the problem of British supremacy that did not limit itself only to trade and the control of the seas, but extended to the affairs of Europe, exerting influence on other nations and claiming the superiority of the British system of government over those of the continent, and, that being so, also the right to lay down the law to other nations. Goudar and Casanova's work, which was enormously successful and ran to at least fourteen editions until 1774,[5] is of special interest because its criticism of Great Britain—"which claims the dominion of the sea, and at present gives law to several great nations"—echoes the same language used by Vattel only six years earlier when he explained that "the famous scheme of the political equilibrium of balance of power" implied that "no power is able absolutely to dominate, or to prescribe laws to others."[6]

Eight years into the future, Casanova would find himself involved in a series of events destined to provide the basis for Chambrier's analysis. In 1772 two Venetian adventurers of Dalmatian origin, Primislao and Stefano Zannovich, who had settled in Amsterdam after being expelled from Venice in 1769,[7] attempted to defraud two merchants originally from Berlin who had also moved to Amsterdam in 1770 and acquired Dutch citizenship. The Zannovich brothers proposed a deal with Chomel and Jordan involving the conversion of a bill of 3500 sequins for 27,000 florins, backed partly by cash and partly by a store of diamonds in Genoa which turned out to be non-existent. Once the fraud was uncovered, Primislao Zannovich promised to indemnify the Dutch merchants and, before fleeing to Naples to join his brother, presented himself to the diplomatic representative of the Republic of Venice in The Hague under the false name of Nicolò Peovich. There he claimed to be the owner of an invented trading company, Peovich & Co., and obtained from the diplomat guarantees in favour of Chomel and Jordan. After this, with the guarantees in hand, he presented himself to Chomel and Jordan and not only convinced them of his solvency by dint of the vaunted diamonds, but also obtained a further loan of 6000 florins. With this money the Zannovich brothers hired a ship, insuring it for 130,000 florins in the United Provinces and Great Britain, and, when it deliberately did not arrive at its destination, they tried to claim compensation. Only then were they finally unmasked and only then did the Venetian diplomatic representative realise the mistakes he had made.[8]

The result was a diplomatic dispute destined to occupy the royal courts of Europe for over ten years, and which involved not only the United Provinces and the Republic of Venice but also the Kingdom of Prussia, the birthplace of Chomel and Jordan. The States General of the United Provinces held the Republic of Venice responsible on the grounds that the Zannovich brothers were its citizens and because its diplomatic representative, Simone Cavalli, had provided the guarantees, but Venice immediately rejected any responsibility. In 1777, after a new attempt to obtain compensation, Chomel asked the Dutch ambassador to Vienna to put pressure on Joseph II and through him on Venice. This led to Chomel being recalled to his home country and put on trial, where, however, he was found innocent. In the meantime, Stefano Zannovich, on the run in Europe, had taken on the false name of Stefano Pastorvecchio, Prince of Montenegro, and in this guise tried to fake his death by publishing, in Dresden (1775), the posthumous works of the deceased Stefano Zannovich.

In the course of 1783 efforts to find a diplomatic solution to the question increased, and Venice decided to approach the Habsburg court to reach a mediated settlement, that is, an agreement between the parties, or, even better, arbitration, that is, a final decision pronounced by a neutral party, such as the emperor. The Dutch rejected all these proposals and on 10 May 1783 the Venetian ambassador to Vienna, Sebastiano Foscarini, was forced to tell the senate that he had failed to convince Kaunitz to take a position on this "thankless affair" and "appalling negotiation." The prudence of Joseph II had by then hardened into a refusal to intervene since "the Dutch do not allow the interposition of foreign courts" and "this agitation has proved disturbing to the tranquillity of commerce."[9]

Left unsatisfied by the recall of the Venetian representative Cavalli and by the responses of the Venetian senate, in January 1784 the States General of the United Provinces ordered retaliation, first by sequestering Venetian ships in Holland and then, since there were none at that time in Dutch ports, affirming the right of Dutch ships in the Mediterranean to seize Venetian ones and give them an armed escort on the route from Malaga to Smyrna.[10] However, Venice still refused to pay and indemnify the Dutch and sought once more to obtain the mediation of the Habsburgs by asking Joseph II to serve as arbitrator. Frederick II of Prussia, for his part, pressured Venice to pay up, while Russia feared that behind all of this was a secret agreement between France and the Dutch party opposed to the Statolder to reinforce, with the pretext of protecting Dutch ships, its military presence in the Mediterranean. All the newspapers of Europe were by then writing about these events and publishing continuous updates.

In October 1784 it appeared that Europe was on the brink of international armed conflict.[11] Each country, beginning with the United Provinces and the Republic of Venice, claimed the right to act according to its own laws. Weapons and warships were being prepared in the arsenals. Public opinion was divided between those who praised the principles of free trade on which the Republic of the Dutch was founded, and those who defended the tradition, antiquity and immutability of Venetian prestige. And, above all, the problem was that no rule in the law of nations appeared to be applicable and able to resolve the question satisfactorily and definitively, in a legal sense, while the Habsburg emperor seemed ill-disposed to exercise his authority to bring about a compromise.

VATTEL IS NOT SUFFICIENT

The problems unleashed on the international scene, but also within the states, were of no small importance. The final one, in chronological terms, was caused by the Dutch refusing to acknowledge the role of the emperor, who the States General eventually agreed to only as a mediator, that is, as someone who proposed a possible solution, while the Venetians instead wanted him to be considered as an arbitrator, that is, as a judge. The Dutch refusal was based on a theoretical question that Vattel had explicitly tackled in the *Droit des gens*. He had, in fact, been one of the first to distinguish clearly between international mediation and international arbitration, underlining the non-binding nature of the former and the binding nature of the latter.[12]

After this, however, the *Droit des gens* had also described a scenario in which arbitrators might arrive at a manifestly unjust decision that was contrary to reason, such as to make them lose all credibility as impartial third parties. Arbitration, therefore, was not a guarantee of impartiality and recourse should only be made to it for questions deemed dubious by both parties and without its decision implying the submission of one sovereign state to another for the reparation of a wrong suffered. Thus, by implication, the *Droit des gens* appeared to agree with the position of the Dutch, or at least raised doubts about the opportuneness of arbitration.

Having failed in its attempt to refer the question to the arbitration of the Emperor, and faced with new pressure from the Dutch, in the spring of 1784 the Republic of Venice set up a special commission composed of a number of senators to re-examine the case in the light of the principles of the law of nations and its compatibility with Venetian law. Once again, from the technical point of view, the problem was on the one hand that of recognising and not violating the prerogatives of the sovereignty of the two states, and on the other that of understanding the international repercussions of a sentence passed by an internal court, in this case the Venetian court that absolved Cavalli, a judgement considered unjust by the other party. The situation thus reversed the Dutch judgement that had found in favour of Chomel and Jordan.

In this case, however, the problem was complicated by the two sentences that followed one another and thus by the question of whether Venice could indirectly decide on a matter already addressed by the Dutch court. As Chambrier would recall in his memoirs of 1794, the *Droit des gens* once again came into play and, referring to Vattel, Chambrier observed that "The Prince should therefore not intervene in cases involving his overseas subjects and grant them protection only in the event of a denial of justice or obvious and palpable injustice, or of a clear violation to the detriment of his foreign subjects in general."[13] Nonetheless, the case of Chomel and Jordan showed that Vattel's statement was "in contradiction with what he argues below, that a foreign Prince may intervene against a sentence handed down to one of his subjects if this is obviously unfair, and if procedures have manifestly been violated."[14]

A series of interpretative problems emerged, which not only explained the Dutch refusal and Joseph II's reluctance to serve as arbitrator, but also—and perhaps for the first time—the difficulty of putting into practice the principles established by Vattel. Moreover, the international context threw doubt on the possibility of Joseph II playing an arbitrator role in the affair. In fact the sovereign, once enthroned after the death of his mother Maria Theresa in 1780, had engaged in a showdown with the United Provinces to question the "Barrier Treaty" signed in Utrecht in 1715, which had accepted the Dutch blockade of navigation on the Scheldt and recognised the United Provinces' right of garrison in Veurne, Ypres, Menen, Tournai and Namur.[15]

During the months of April and May 1784 it had therefore become clear to Venice that the Chomel and Jordan question could not be resolved either through international diplomacy or through the law of nations. It had now entered the arena of European public opinion, and all the newspapers were writing about it, and even Simon-Nicolas-Henri Linguet, in his famous *Annales politiques, civile set littéraires du dix-huitième siècle*, announced that he would address it more fully, since

> we can deceive ourselves that these two sisters will not give Europe the scandal of seeing them slit each others' throats and that in a time so unfavorable to republics, those that have the good fortune of continuing to exist will not compromise their existence by fighting or have the happiness of existing ecore will not compromise their own existence by fighting or derision.[16]

Linguet's decision, or more accurately his threat, to deal publicly with the dispute between Venice and the United Provinces risked undesirable repercussions. Between 1777 and 1787 he had been publishing a series of articles in the *Annales politiques* on the very topic of the law of nations, supporting principles that went directly against Vattel's thesis. He claimed that there was no other natural law of nations than that of the strongest, that war was itself a natural and necessary right—indeed "the most active of rights"—which meant that violence was the foundation of public law.[17]

Linguet thus went against all those who, starting with the *Encyclopédistes*, argued for Europe's moral superiority on the grounds of its ability to use reason as a means of deterring recourse to war and of pursuing peace. Opposed to any form of mediation, arbitration or international tribunal, Linguet maintained that only the courage and strength of the parties could resolve a confrontation between two or more states. For him, not even commerce was to be considered an instrument of peace, but rather an opportunity to put into practice the warrior virtues of the nations.

In other words, Linguet's position was one that promoted a new concept of politics of power which stipulated that in international relations the law of nations and politics as a moral science had to be completely excluded in favour of an idea of reason of state placed not at the service of the glory of the king but at the service of the interests of the peoples.[18] Needless to say, this approach was diametrically at odds with any notion of the "spirit of commerce" being an instrument of "good government" and also against any hypothesis of republicanism—whether the modern version of the Dutch or the ancient one of the Venetians—embodying the principles of a good government of the nations. In fact, Linguet had already explicitly expressed his opposition to this idea of "good government" in his *Annales*, enjoining the reader not to believe that such systems made reasons for conflict disappear.[19]

As intimated, Linguet "was too famous and important to be answered only with silence."[20] The Republic of Venice thus decided to fight the battle with other arms, those of literary polemics. In February 1784 the ambassador of Venice to Vienna, Foscarini, hired Casanova, who had already shown his worth to the Republic as a spy and as an apologist in his role as co-author of *The Chinese Spy* and author of the *Confutation* of Amelot de la Houssaye's *Histoire du gouvernement de Venise*.[21] This was an unofficial assignment, but Foscarini—probably using money that came from the Venetian government itself, given that his own economic resources were limited—not only gave Casanova access to all the documents he needed to respond to the Dutch but also the means to print a series of pamphlets, which appeared—some under a false name, some anonymously—in Germany, Venice and Holland itself.

CASANOVA A READER OF VATTEL?

From the vast critical literature on Casanova we learn that he completed his studies in jurisprudence in Padua, where he graduated in 1742, probably having followed the lessons of certain philosophers including Jacopo Stellini, but preferring Gassendi in particular.[22] In Paris in 1752 and 1753 and in the years immediately afterwards he had familiarised himself with clandestine literature, especially Toland, Freret, Collins, Boulanger, La Mettrie and Helvétius. In most cases his knowledge of the work of these authors was not direct, but came via summaries and manuscripts. Casanova himself, in the *Confutazione della Storia del governo veneto* and in *Supplimento*, had referred to authors and texts that he had obtained second hand, most importantly Bayle's *Dictionnaire*.[23] He was an admirer but jealous of Voltaire, leaving behind some entertaining accounts of encounters with him,[24] and was a student of the works of Bernardin de Saint-Pierre, particularly the *Examen des Études de la nature et de Paul et Virginie* (1788–1789) from which emerged—again probably through other authors—references to Wolff and Hutcheson.[25]

As can be gleaned from these brief references, for years Casanova frequented diplomatic circles, was interested in international politics and events, and wrote about them, albeit without using his legal education or the scientific literature of the time in an analytical way. His approach to these subjects was essentially literary, practical and polemic, and was implemented through rhetoric and strategies aimed at capturing the attention of the general public, rather than that of rulers or those involved in the administration of justice or international arbitrage.

This was precisely why Casanova was employed in the dispute between Venice and the United Provinces. In Vienna, thanks to Ambassador Foscarini, he stayed for almost two years until March 1785, the time of the ambassador's death, to write a series of works on the Zannovich/Chomel and Jordan controversy in favour of Venice.[26] Although this history has not yet been fully reconstructed, it does help us to understand how an international controversy was handled through a kind of "parallel diplomacy"—of literature and public opinion—since the principles of the law of nations enunciated by Vattel and before him Wolff and Grotius were difficult to apply in the circumstances.

Having arrived in Venice on 18 February 1784 after a journey to Dresden, Berlin, Dux and Dessau, Casanova found Foscarini (an old acquaintance and a friend of Casanova's protector, the Venetian senator Pietro Zaguri) in the midst of a full-blown crisis over the affair between Holland and Venice because the States General had decided on 9 January 1784 to seize all Venetian ships in Dutch waters. Casanova—as he would recount in his *Histoire de ma vie*—had personally been witness to the birth of the 1772 scam (he had played cards with Primislao Zannowich and Lord Lincoln)[27] and knew the protagonists. One can therefore see why he immediately offered, or was asked to provide, his services as a libelist to Foscarini so as to refute the Dutch and contest, in the first instance, some articles that had appeared in Elie Luzac's *Gazette de Leyde*.[28]

Casanova worked quickly: one of his initial Italian manuscripts was translated into French by his friend Heinrich Wolfgang Behrisch (known as *chevalier de Béris*) under the title *Lettre historico-critique sur un fait connu, dependant d'une cause peu connue, addressée au Duc de **** before 12 March 1784, and submitted to the Austrian censors on 19 March. Owing to difficulties posed by the censorship, the work was then sent to Dessau's *Libraire des Savans* (headed by Behrisch's brother), where it was published on 12 May. According to documents of the time, Casanova had paid the press directly, no doubt with money given by Foscarini. Two manuscript versions of this work survive among Casanova's papers in the archives of the Muzeum města Duchcova, while extracts in German and French translation appeared in certain European publications. Another version was serialised in Italian between 22 January and 26 February 1785 in the newspaper of the free port of Trieste, *Osservatore Triestino*, and in this case the place of publication was probably chosen because Trieste was the main Austrian port of the Adriatic, and because it was close to Venice.[29]

Immediately afterwards Casanova went on to publish l'*Exposition raisonnée du different qui existe entre les deux Républiques de Venise et d'Hollande*, a seventy-nine-page libretto that bore the date 1785. Research has revealed that the drafting of the text dates back to the period after August 1784, and therefore Casanova must have dedicated the whole of early summer to collecting and copying the documents needed from the records office of the Venetian embassy in Vienna. Around the same time, *Exposition* was also published in Italian, in two different editions entitled

Esposizione ragionata della contestazione che sussiste tra le due Repubbliche di Venezia e di Olanda.[30]

But the literary dispute was not destined to end there. In the numbers 21 and 22—of 15 March and 18 March 1785—of the *Gazette de Leyde*[31] the editors, Jean and Etienne Luzac, published the Dutch point of view on the affair, and Casanova responded to them with another pamphlet dated 31 March and entitled *Lettre à Messieurs Jean et Etienne L., contenant des observations sur le Narre de l'affaire, quia donné leu au différent entre la République de Venise & celle d'Hollande*, which was immediately translated into Italian and Dutch.[32] Finally, a few weeks later Casanova also published *Supplément à l'exposition raisonnée du different qui subsiste entre la République de Venise & celle de Hollande*, another libretto, this time of ninety-six pages, also published in an Italian edition, probably in the month of April and different from the *Exposition* in so far as the narrative of the text was essentially a chronicle, and more space was given to the argumentative thrust and to legal reflections.[33] But in the meantime, on 23 April 1785, Ambassador Foscarini had died and with him Casanova's main source of funding.

JUDGES AND MEDIATORS

As mentioned, there is a notable difference in the tone of argument between Casanova's first pamphlet, entitled *Exposition raisonnée*, and the later *Supplément* one. The first text was built around the need to impress the reader and convince him of Venice's valid stance by exposing the alleged press and international political campaign against the republic, which was a result of excessive press freedom and the penchant for conspiracy theories (*le cabale*) dreamt up in cafés. Casanova, the author, concentrated on the reconstruction of facts and on particular details likely to arouse the curiosity of the public, such as the forging of documents, the character of the people involved (Zannovich's "tears of a conman"), the existence and the disappearance of diamonds, and the clandestine trade of 3000 barrels of wine between Dalmatia and the United Provinces.

In the *Supplément* instead Casanova abandoned the literary form to develop a logical text of systematic reasoning addressed as a letter to Linguet. The recipient and the target were therefore different and clearly identified, and Casanova knew that he could not compete with Linguet on strictly literary ground. The argument was therefore constructed analytically, in the form of the publication, with commentary, of a series of documents, produced mainly by the Republic of Venice, aimed at convincing the reader not so much by the development of the facts but, in Casanova's words, by the "authenticity of the exposition."

The narration of the facts was thus gradually replaced with legal argumentation and it may be supposed that just as Foscarini had supplied Casanova with the documents, so he had also helped provide the arguments. It should therefore not surprise us that in confronting Linguet— who, as we have seen, was a supporter of the politics of power who held the conventional law of nations to be useless—Casanova's observations seemed instead to echo the words of Vattel based on the principle of formal equality between nations and on the natural foundation of the conventional law of nations. For example, regarding the role played by Cavalli, the Venetian diplomatic representative accused by the Dutch of having provided guarantees and help to the Zannovich brothers was tried in his home country and found not guilty. Casanova explained that Cavalli's innocence derived from his status as a minister and a diplomat, which made it impossible to declare him personally liable for the money defrauded from Chomel and Jordan.[34]

These were words that appeared to restate the passage in the *Droit des gens* where Vattel recalled that the inviolability of an overseas minister determined his total exemption from the jurisdiction of the state in which he resided[35] and that this dispensation did not depend only on a simple convention between nations but derived from natural law itself. Second, to the Dutch accusation—reiterated by Linguet—of wanting to help Cavalli escape his responsibilities, Casanova replied that in fact the very opposite was true and that Venice, using as a basis the principle of authorised jurisdiction for diplomatic agents and the ambassador's exemption from the civil jurisdiction of the host state, had promptly set up an extraordinary criminal court,[36] thereby respecting the previously mentioned principle recalled by Cornelis van Bynckershoek in *De foro legatorum* (1721) and also by Vattel.[37]

Among the various other arguments drawn from the law of nations, special place was given over to the question of arbitration and the role that the emperor could and should have played between the two disputants. As noted, the issue was that of choosing between mediation and arbitration, and Holland had not accepted the emperor as arbitrator because the question was not likely to be resolved by a binding decision between the parties, nor could it be overcome by mediation alone. Casanova observed that the emperor could rightfully accept only the role of arbitrator because that

was the one most compatible with his royal status.[38] "The task of a mediator," Casanova wrote, "is to temporise so as to reach an agreement
between the parties, while that of an arbitrator is to examine the facts and
after having weighed the explanations of both sides to pronounce a final
judgment."

These words were not dissimilar to Vattel's, according to whom "mediation, in which a common friend interposes his good offices, frequently
proves efficacious in engaging the contending parties to meet each other
half-way,—to come to a good understanding,—to enter into an agreement or compromise respecting their rights,—and, if the question relates
to an injury, to offer and accept a reasonable satisfaction." Conversely, in
arbitration, "When once the contending parties have entered into articles
of arbitration, they are bound to abide by the sentence of the arbitrators:
they have engaged to do this; and the faith of treaties should be religiously
observed."[39]

Casanova's pamphlets on the dispute between Venice and Holland
should be read in the light of Amelot de la Houssaye's *Confutazione della
storia del governo,* published fifteen years earlier. In that work the problem
was the defence of the good government of the Republic of Venice not
against Amelot (who had died in 1706), but against French "despotism"
that turned into cultural supremacy. The real interlocutors of the
Confutazione had in fact been, polemically, D'Alembert and Diderot[40]
with their view of the Republic transmitted through the relative entry in
the *Encyclopédie* and the invitation contained there to read Amelot. In the
mid-1780s, Casanova resumed the same rhetorical strategy to defend,
against the legal arguments of the *Droit des gens,* the good government of
Venice threatened by Dutch despotism and their arrogant trade policy.

The sudden death of Ambassador Foscarini on 23 April 1785 momen-
tarily deprived Venice of a diplomatic representative in Vienna and
Casanova of a protector who made use of his pen to explain to the
European public the position of the Republic of Venice. In July 1785,
however, the new Venetian representative to The Hague informed the
senate that the States General intended for the moment to leave the matter open and to not persist with their claims,[41] and in September of the
same year Casanova left Vienna to settle in Chateaux Duchcov, Bohemia,
where he held the post of librarian of the Count of Waldstein until his
death. Pierre Chomel, however, continued for many years to seek justice
for the fraud perpetrated against him and in 1797, twenty years after the
start of the *affaire,* asked Napoleon Bonaparte for help to make his claim,[42]

before the final traces disappeared and with them all memory of an international case and the resultant clash between the two most famous republics in Europe.

NOTES

1. Jean-Pierre de Chambrier, "Questions de droit des gens et Observations sur le Traité du Droit des Gens de M. de Vattel," in *Mémoires de l'Académie royale des Sciences et Belles-lettres depuis l'avénement de Fréderic Guillaume II au throne, MDCCLXXXVIII et MDCCLXXXIX* (Berlin: Decker, 1793), 436–492.

2. Vincent Chetail, "Vattel et la sémantique du droit des gens: une tentative de reconstruction critique," in *Vattel's International Law in a XXIst Century Perspective / Le droit International de Vattel vu du XXIe siècle*, ed. Vincent Chetail and Peter Haggenmacher (The Hague: Martinus Nijhoff, 2011), 419–420.

3. ASV, Inquisitori di Stato, busta 177, nr. 824 letter of the Inquisitori di Stato to the Ambassador Sebastiano Foscarini, Vienna, 30 March 1782.

4. Sophus A. Reinert, *Translating Empire. Emulation and the Origins of Political Economy* (Cambridge, MA: Harvard University Press, 2011), 13–16, 295. On Casanova as a journalist see Marie-Françoise Luna, "Giacomo Casanova de Seingalt (1725–1798)," in *Dictionnaire des journalistes (1600–1789)*, ed. Anne-Marie Mercier-Faivre and Denis Reinaud (2010), http://dictionnaire-journalistes.gazettes18e.fr/journaliste/143-giacomo-casanova-de-seingalt

5. Francis L. Mars, "Ange Goudar, cet inconnu (1708–1791)," *Casanova Gleanings* IX (1966), 25–28; James R. Childs, *Casanoviana. An Annotated World Bibliography* (Vienna: Nebehay, 1956), 12–13.

6. Reinert, *Translating Empire*, 13, 15 with reference to Ange Goudar, *The Chinese Spy* (London: Bladen, 1765), vol. 4, I and Vattel, *Law of nations* (Dublin: White, 1787), 468.

7. On Stefano Zannovich see Franco Venturi, *Settecento riformatore*, 5/2, *L'Italia dei Lumi. La Repubblica di Venezia* (Turin: Einaudi, 1990) 358–360; Roland Mortier, *Le "prince d'Albanie". Un aventurier au siècle des Lumières* (Paris: Honoré Champion, 2000).

8. A brief history of this *affaire*, based on Dutch documents, is offered by da E.O.G. Haitsma Mulier, "De affaire Zanovich. Amsterdams-Venetiaansche betrekkingen aan het einde van de achttiende eeuw," *Amstelodamum* 72 (1980): 85–119.

9. ASV, Senato, Dispacci degli ambasciatori, Germania, busta 285, letter of Foscarini, Vienna, 10 May 1783, cc. 77–81.

10. Koen Stapelbroek and Antonio Trampus, "Commercial reform against the tide: Reapproaching the eighteenth-century decline of the republics of Venice and the United Provinces," *History of European Ideas* 26 (2010), 192–202.

11. Franco Venturi, *Settecento riformatore*, 4/2, *La caduta dell'Antico Regime (1776–1789)* (Turin: Einaudi, 1984), 583–587.

12. Vattel, *Droit des gens*, book II, ch. XVIII, § 328–329.

13. Jean-Pierre de Chambrier, "Questions de droit des gens," 444, with reference to Vattel, *Droit des gens*, book II, ch. VIII, § 108.

14. Chambrier, "Questions de droit des gens," 446.

15. Marc Belissa, *Fraternité universelle et intérêt national (1713–1795). Les cosmopolitiques du droit des gens* (Paris: Kimé, 1998), 124; Stapelbroek and Trampus, "Commercial reform," 200–201.

16. Nicolas-Henri-Samuel Linguet, *Annales politiques, civiles et littéraires du dix-huitième siècle* (London: Spilsbury and Snowhill, 1784), vol. 11, 60.

17. Belissa, *Fraternité universelle et intérêt national*, 40–41.

18. Belissa, *Fraternité universelle et intérêt national*, 99.

19. See Nicolas-Henri-Samuel Linguet, ed. *Mélanges de politique et de littérature* (Boullon: n.p., 1778), 348–349.

20. Venturi, *Settecento riformatore*, 4/2, *La caduta dell'Antico Regime*, 586.

21. Giacomo Casanova, *Confutazione della storia del Governo Veneto d'Amelot de la Houssaie. Divisa in tré parti. Parte prima [–seconda]*, with *Supplimento all'opera intitolata Confutazione della storia del Governo Veneto d'Amelot de la Houssaie* (Amsterdam [Lugano]: Mortier [Agnelli], 1769). The *Histoire* by Amelot de la Houssaye was published in Paris, by Frédéric Leonard, in 1676.

22. Tom Vitelli, "Casanova and Gassendi. Proposition for a study," *Casanova Gleanings* 1980, 11–14; Giacomo Casanova, *Pensieri libertini*, ed. Federico Di Trocchio (Milan: Rusconi, 1990), 25–27.

23. Casanova, *Pensieri libertini*, 30.

24. Ted Emery, "Casanova's Coffeehouse: Sociability, Social Class, and the Well-bred Reader in Histoire de ma vie," in *The Thinking Space. The Café as a Cultural Institution in Paris, Italy and Vienna*, ed. Leona Rittner, W. Scott Haine and Jeffrey H. Jackson (London-New York: Routledge, 2016), 180–181.

25. The manuscript preserved in Dux/Duchcov was not published until the twentieth century: Jacques Casanova de Seingalt, *Examen des Études de la nature et de Paul et Virginie de Bernardin de Saint-Pierre*, ed. Tom Vitelli and Marco Leeflang (Utrecht: L'Intermédiaire des Casanovistes, 1985), 45.

26. Casanova wrote, in the *Précis de ma vie* (dating back to 1788/1789 and published for the first time in Jacques Casanova, *Le Messager de Thalie-Le*

Précis de ma Vie, Paris: Fort, 1925, 144): "je me suis placé au service de M. Foscarini, ambassadeur de Venise, pour lui écrire la dépêche."

27. Giacomo Casanova, *Histoire de ma vie*, vol. III, ed. Gérard Lahouati and Marie-Françoise Luna (Paris: Gallimard, 2015) 819.

28. On Luzac's *Gazette de Leyde* see Jeremy D. Popkin, *News and Politics in the Age of Revolution: Jean Luzac's Gazette de Leyde* (Cornell: Cornell University Press, 1989).

29. ASV, Archivio proprio dell'ambasciata in Germania a Vienna, busta 145, Lettere missive e responsive 1780–1786.

30. Childs, *Casanoviana*, 72–75.

31. Popkin, *News and Politics in the Age of Revolution*, 222.

32. Childs, *Casanoviana*, 75–78.

33. Childs, *Casanoviana*, 79–80.

34. Giacomo Casanova, *Supplimento alla esposizione ragionata della controversia che sussiste tra la Repubblica di Venezia e quella d'Olanda* ([Venice]: n.p., 1785), IX.

35. Vattel, *Droit des gens*, book IV, ch. VII, § 92.

36. Casanova, *Supplemento*, XIII.

37. Vattel, *Droit des gens*, book IV, ch. VIII, § 110.

38. Casanova, *Supplemento*, XLIII.

39. Vattel, *Droit des gens*, book II, ch. XVII, § 328–329.

40. Branko Aleksić, "Casanova et D'Alembert," *Recherches sur Diderot et l'Encyclopédie* 42 (2007), 83–94, https://journals.openedition.org/rde/2343

41. Haitsma Mulier, "De affaire Zanovich," 114.

42. *Correspondance inédite officielle et confidentielle de Napoleon Bonaparte-Venise* (Paris: Panckoucke, 1819), 183–193.

BIBLIOGRAPHY

Most of the quotations of Vattel's *Droit des gens* come from the widely available English-language edition of 2008, which includes an introduction by Béla Kapossy and Richard Whatmore (Indianapolis: Liberty Fund) and maintains the text and English title *The Law of Nations: Or, Principles of the Law of Nature Applied to the Conduct and Affairs of Nations and Sovereigns* of the 1797 London standard edition.

Contemporary translations of primary sources are listed under the names of their authors. Full manuscript sources are referenced only in the notes for reasons of space.

ARCHIVAL ABBREVIATIONS

ASV Archivio di Stato di Venezia, Venice, Italy

PRIMARY SOURCES

PRINTED BOOKS

Casanova, Giacomo, *Confutazione della storia del Governo Veneto d'Amelot de la Houssaie. Divisa in tré parti. Parte prima [-seconda]*, with *Supplimento all'opera intitolata Confutazione della storia del Governo Veneto d'Amelot de la Houssaie* (Amsterdam [Lugano]: Mortier [Agnelli], 1769)

———, *Examen des Études de la nature et Paul et Virginie de Bernardin de Saint-Pierre en 1788*, eds. Tom Vitelli and Marco Leeflang (Utrecht: L'Intermédiaire des Casanovistes, 1985)

———, *Exposition raisonné du différent, qui subsiste entre les deux Républiques de Venise et d'Hollande* ([Venise]: n.p., 1785; Italian translation *Esposizione ragionata della contestazione che sussiste tra le due Repubbliche di Venezia e d'Olanda* [Venezia]: n.p., 1785).

———, *Histoire de ma vie*, vol. III, eds. Gérard Lahouati and Marie-Françoise Luna (Paris: Gallimard, 2015)

———, *Le Messager de Thalie-Le Précis de ma Vie* (Paris: Fort, 1925)

———, *Pensieri libertini*, ed., Federico Di Trocchio (Milan: Rusconi, 1990)

———, *Supplément à l'exposition raisonné du différend qui subsiste entre les deux Républiques de Venise et d'Hollande* ([Venise], n.p. 1785; Italian translation *Supplimento alla esposizione ragionata della controversia che sussiste tra la Repubblica di Venezia e quella d'Olanda* ([Venice]: n.p., 1785)

Chambrier, Jean-Pierre de, "Questions de droit des gens et Observations sur le Traité du Droit des Gens de M. de Vattel," in *Mémoires de l'Académie royale des Sciences et Belles-lettres depuis l'avénement de Fréderic Guillaume II au throne, MDCCLXXXVIII et MDCCLXXXIX* (Berlin: Decker, 1793), 436–492

Correspondance inédite officielle et confidentielle de Napoleon Bonaparte-Venise (Paris: Panckoucke, 1819)

Goudar, Ange, *The Chinese Spy* (London: Bladen, 1765)

Linguet, Nicolas-Henry-Samuel, *Annales politiques, civiles et littéraires du dix-huitième siècle* vol. 11 (London: Spilsbury and Snowhill, 1784)

———, ed., *Mélanges de politique et de littérature* (Bouillon: n.p., 1778)

Vattel, Emer de, *Law of Nations: Or, Principles of the Law of Nature, Applied to The Conduct and Affairs of Nations and Sovereigns*, eds. Béla Kapossy and Richard Whatmore (Indianapolis: Liberty Fund 2008)

SECONDARY SOURCES

Aleksić, Branko, "Casanova et D'Alembert," *Recherches sur Diderot et l'Encyclopédie* 42 (2007), 83–94

Belissa, Marc, *Fraternité universelle et intérêt national (1713–1795). Les cosmopolitiques du droit des gens* (Paris: Kimé, 1998)

Chetail, Vincent, "Vattel et la sémantique du droit des gens: une tentative de reconstruction critique," in *Vattel's International Law in a XXIst Century Perspective / Le droit International de Vattel vu du XXIe siècle*, eds. Vincent Chetail and Peter Haggenmacher (The Hague: Martinus Nijhoff, 2011), 385-434

Childs, James R., *Casanoviana. An Annotated World Bibliography* (Vienna: Nebehay, 1956)

Emery, Ted, "Casanova's Coffeehouse: Sociability, Social Class, and the Well-bred Reader in Histoire de ma vie," in *The Thinking Space. The Café as a Cultural Institution in Paris, Italy and Vienna*, eds. Leona Rittner, W. Scott Haine and Jeffrey H. Jackson (London-New York: Routledge, 2016), 169–184

Haitsma Mulier, E. O. G., "De affaire Zanovich. Amsterdams-Venetiaansche betrekkingen aan het einde van de achttiende eeuw," *Amstelodamum* 72 (1980), 85–119

Luna, Marie-Françoise, "Giacomo Casanova de Seingalt (1725–1798)," in *Dictionnaire des journalistes (1600–1789)*, eds. Anne-Marie Mercier-Faivre and Denis Reinaud (2010) (http://dictionnaire-journalistes.gazettes18e.fr/journaliste/143-giacomo-casanova-de-seingalt).

Mars, Francis L., "Ange Goudar, cet inconnu (1708–1791)", *Casanova Gleanings* IX (1966), 43–44

Mortier, Roland, *Le "prince d'Albanie". Un aventurier au siècle des Lumières* (Paris: Honoré Champion, 2000)

Popkin, Jeremy D., *News and Politics in the Age of Revolution: Jean Luzac's Gazette de Leyde* (Cornell: Cornell University Press, 1989)

Reinert, Sophus A., *Translating Empire. Emulation and the Origins of Political Economy* (Cambridge, Mass.: Harvard University Press, 2011)

Stapelbroek, Koen, and Trampus, Antonio, "*Commercial reform against the tide: Reapproaching the eighteenth-century decline of the republics of Venice and the United Provinces*" in *History of European Ideas*, vol. 36 (2010), 192–202

Venturi, Franco, *Settecento riformatore*, vol. 5/2, *L'Italia dei Lumi (1764–1790). La Repubblica di Venezia (1761–1797)* (Turin: Einaudi, 1990)

———, *Settecento riformatore* vol. 4/2 *La caduta dell'Antico Regime 1776–1789* (Turin: Einaudi, 1984)

Vitelli, Tom, "Casanova and Gassendi. Proposition for a study," *Casanova Gleanings* 1980, 11–14

CHAPTER 9

From Natural Rights to the Rights of Man

In recent years the attention of historians of the modern age has returned to reflect on the problem of human rights and their origin, both because of their continual violation and because of the cultural debates that have called into question the centrality of the West in the historical genesis and in the construction of a modern theory of the rights of man. This discussion, which is still ongoing, also sees at the centre of the historical debate the problem of the relevance and autonomy of the culture of the Enlightenment. In other words, the debate is trying to understand whether the invention of human rights was a typical product of eighteenth-century Enlightenment or whether it has older roots traceable in a long tradition that, running from antiquity, through Christianity and Humanism, has traversed Western history itself.[1]

The problem is complex also because it forces us to investigate and understand the function of natural law in identifying natural rights and in transforming them, within the culture of the *ancien régime*, into a science of duties and rights useful to the most modern theories of human rights. The subject of rights and duties was in fact the object of the reflection of many protagonists of the culture of the seventeenth and eighteenth centuries, from Grotius to Barbeyrac, from Locke to Pufendorf, up to Wolff and Burlamaqui. With the *Droit des gens* Vattel took his place in this history of human rights at a vital moment corresponding to an

Archival Abbreviations
ASM   Archivio di Stato di Modena, Modena, Italy

© The Author(s) 2020
A. Trampus, *Emer de Vattel and the Politics of Good Government*,
https://doi.org/10.1007/978-3-030-48024-0_9

155

epistemological rift after which Enlightenment culture transformed the natural rights of man into moral and political rights, that is, accentuating their political character and placing them within constitutions so as to make it easier to defend and protect them during the crisis of the *ancien régime* and the political and economic changes of the late eighteenth century.

THE SPIRIT OF RIGHTS

The history of the reception of the *Droit des gens* therefore represents a useful instrument for understanding the birth of the modern theory of human rights, both because Vattel is recognised as an important author in the synthesis of the natural law tradition that preceded him, and because he in turn became one of the key reference points for intellectuals and philosophers in the 1780s who were occupied with rebuilding the science of human rights to make it the essential basis of declarations of rights and of the earliest written constitutions.

One of the most significant authors from this point of view, as well as for his ability to rework Vattel's teaching and for the influence of his own work, was Gaetano Filangieri, the Neapolitan philosopher who was admired by Benjamin Franklin and convinced of the possibility of transforming natural equality into an equality of rights by placing man with his rights to freedom and property at the centre of his reflection.[2] Filangieri's 1780 work, *Science of Legislation*, provides a highly interesting case study. Not only did it overcome geographical variations by being widespread through the Mediterranean area, continental Europe and the Americas, but it also challenged temporal categories: in fact, modern concepts of constitutionalism and human rights—still vital in the political cultures of the nineteenth and twentieth centuries—stemmed from Filangieri's work.

As shown by the mapping of the times, places and frequency of editions and translations from 1780 to 1864, the *Scienza della legislazione* was a publishing and cultural phenomenon from which it is possible to reconstruct a variety of textual appropriation strategies in different cultures and ages. Since the publication of the first Italian edition, and despite its length (seven volumes), the *Scienza della legislazione* had become an Enlightenment bestseller in Europe. The success of the work was also magnified by the heroic end of the author, who died aged 35 before the French Revolution, in 1787. Many other Enlightenment masterpieces enjoyed widespread circulation in part because of their small typographical size (e.g. Cesare Beccaria's *Dei delitti e delle pene*). Conversely, the *Scienza della legislazione* required enormous effort as far as the translation was concerned, as well as huge financial resources for its publication, which frequently came from a blend of diplomatic, commercial and Masonic networks. During the nineteenth century, when modern constitutionalism was surpassed by newly emerging codes, Filangieri's project no longer seemed appealing, and thus new strategies of textual appropriation were developed. As a result, the text no longer circulated in its entirety but in parts (dealing with, for example, economics, criminal law or educational reform), in order to answer more effectively the needs of various post-Restoration national and cultural contexts.

The *Scienza della legislazione*, along with its various editions, translations and commentaries, allows us to understand how much of Enlightenment Europe moved through the Revolutionary and Napoleonic years and flowed into the new Europe of the nineteenth century. Filangieri's thought and republican heritage were not swallowed whole by the French Revolution, but were infused, redefined and reinterpreted by nineteenth-century liberal culture. They were adapted to respond to new social needs and to the political questions that emerged during the Restoration. The *Scienza della legislazione* should therefore not be considered as a work that remained static and fossilised in the Enlightenment, since in fact it was able to overcome temporal categories and extend its influence, beyond the Mediterranean area and the Enlightenment. This process, extraordinary in some of its facets, cannot be fully understood without an analysis of the seeds of modernity contained in the *Scienza della legislazione*. In fact, it was a mammoth constitutional project that engaged philosophy of law, economics and commerce, criminology and the system of crimes and punishments, upbringing, public education, and religion. It was not just a theoretical text, but was closely connected to late eighteenth-century economic, legal and social transformations. The young author was certainly influenced by the context of his city, Naples, which was still dominated by political backwardness, but it was exactly there and in the Mediterranean area more generally that the survival of rural feudalism clashed with the development of commerce and where the many contradictions of the Enlightenment emerged: the aspiration to equality and universal fraternity against the inequality generated by the growing market economy.

The *Scienza della legislazione* examined these issues in depth, and they were destined to become even more central during the nineteenth and twentieth centuries. Filangieri debated them using modern language and clear words that would stir the reader's emotions and, more concretely, could be easily translated and understood in other European languages. As a result, the newness of the problems discussed by the *Scienza della legislazione* and the modernity of its language are two important elements which explain the strategies of circulation, translation and adaptation of the text across the West in a trans-temporal dimension.

The story of Gaetano Filangieri's work provides at least three key elements that make it a relevant case study for broadening the perspectives of global Enlightenment. Firstly, moving in the relatively classic field of intellectual history, in terms of geography *Scienza della legislazione* circulated in concentric circles: it originated in late eighteenth-century Naples and then reached all of Europe and the Atlantic world. Secondly, it spread in time for two centuries after its publication. Thirdly, it survived beyond its own era, thus reflecting the category of trans-temporality. It did not in fact ossify but was continuously modified in response to different national strategies of appropriation, which used either the whole work or the parts most fitting to specific constitutional and cultural contexts.[3]

The grounds on which Filangieri built his theory of human rights comprised above all the reflection of the American Revolution, its echoes in Europe, the condemnation of slavery in the Atlantic world, and the denunciation of despotism and the absence of freedom. Alongside this, the other area in which Filangieri moved was that of the new inequalities arising from the development of commerce and nascent globalisation. It was there that the new rights of man intertwined, in his way of seeing things, with the rights of liberty and property and with those deriving from free trade, but the crucial problem became that of harmonising the development of the free market with the principles of ethics and of sociability and solidarity between individuals. Filangieri's opposition to Adam Smith was clear. In an international context dominated by the jealousy of trade and rivalries between nations (expressions that derived, as we have seen, from Hume[4]), Filangieri denounced the dangers arising from Smith's ideas, especially where the Scottish economist explained that during the formation of the market wealth distribution occurred naturally without the need for government intervention. Filangieri instead reminded his readers that this "natural" distribution did not necessarily happen in a just and honest manner (expressions that came from Antonio Genovesi) and maintained that it was therefore necessary for politics and good laws to intervene to mitigate inequalities and guarantee everyone the exercise of rights.[5]

When writing the *Scienza della legislazione*, Filangieri did not declare all his sources and, in keeping with the custom of the period, preferred to cite ancient authors rather than contemporary ones. The fact that the *Scienza* did not quote Vattel directly should therefore not come as a surprise, despite the Swiss jurist being a keystone of Filangieri's discourse, especially when the Neapolitan philosopher proposed the constitutional state with its fundamental laws as a guarantor of the rights of man. The reference to Vattel is also apparent in Filangieri's reasoning on the difference between the absolute goodness and relative goodness of laws,[6] as well as in other parts of the *Scienza della legislazione*, primarily in those through which he explained how human rights can be activated, that is, by reforming the system of crimes and punishments and the relationship between the citizen, the constitution and sovereignty.[7] Finally, it is interesting that one of the main protagonists in the dissemination of the *Scienza della legislation* in the Italian peninsula, through the first Venetian reprint of 1782, was Lodovico Antonio Loschi, who only a year earlier—as we have seen in the preceding chapter—had completed his translation of the *Droit des gens*.

The Good Government of Appiano Buonafede

The name of Loschi is connected not only to Vattel and Filangieri, but also to another widely read author in Italian and French culture of the second half of the eighteenth century, namely Appiano Buonafede.

During the 1770s and 1780s, in part as a consequence of the American Revolution, the discussion around Vattel's work in Catholic countries underwent an unexpected change of register. The *Droit des gens*, in fact, found itself at the centre of the argument between the Protestant and Catholic interpretations of natural law. In order to justify criticisms of Vattel's theories, the *Droit des gens* was increasingly classified as the work of a Protestant author and dangerous in the Catholic sphere due to the use that could be made of it through the re-elaboration of the right to resistance in the light of what was happening in the American colonies.

One part of the Catholic culture, which the historiography of the twentieth century came to define as the "Catholic Aufklärung," or "Catholic Enlightenment,"[8] nevertheless undertook to keep alive the dialogue between Protestant and Catholic natural law, and so continued to circulate the works of authors like Grotius, Wolff and Vattel himself.

The main protagonist in this period was undoubtedly the abbot Appiano Buonafede, known as a philosopher and man of letters but also the author of legal and political works destined to be widely circulated in the Italian peninsula as well as to be relatively successful in the German-speaking world.[9] Born in 1716 in Comacchio, a town close to the Po delta and then part of the Papal States, Buonafede studied philosophy in Naples and theology in Rome, and in 1740 became a reader in theology in Naples where he came to know the famous Celestino Galiani, of whom he wrote a biography. In Naples he attended Vico's lectures, became a friend of Antonio Genovesi and breathed the cultural air of Naples that was open to 'modern' authors. After 1755 he moved to Bologna, an enclave of the Papal States, and became close to the Academy of Sciences established by the scientist Luigi Ferdinando Marsili and to the Academy of Arcadia, in which he took the name Agatopisto Cromaziano that he later used in many of his works.

Buonafede, critical of the ancient tradition of scholastic philosophy, was always an attentive reader of the writings of Locke, Newton, Descartes and authors representing the school of modern natural law in general. He himself acquired celebrity for his *Istoria critica e filosofica del suicidio* (1761),[10] which was also translated into French, and for various writings on literary issues, which facilitated his transfer to Rome in 1771 and his career within the religious order of the Celestines. He eventually became procurator general of that order, prefect general (1777), perpetual abbot of San Eusebio (1781) and, finally, from 1791, apostolic vicar in the Papal States.

Buonafede, who was a personal friend of Loschi's, had already written about Vattel in his 1763 work on the *Conquiste celebri esaminate col natu-rale diritto delle genti*, published in the wake of the Seven Years' War. The text used this important military and international political event as a start-ing point to call into question all the principal authors of modern natural law and their theories on just war, on the function of the law of nations, and, more generally, on the relationship between law and morality in the context of relations between peoples and countries.[11] In the *Conquiste* Buonafede examined the opinions of the most authoritative 'modern' political writers regarding the right of conquest (Machiavelli, Grotius, Hobbes, Spinoza, Locke, Montesquieu, Helvétius and Cocceji) from a perspective that centred on the Catholic view of 'reason' and 'humanity.' However, this approach remained open to all Reformist natural law in the hope of possibly 'Catholicising' Protestant natural law both to try to influence it and to modernise Catholic natural law itself with respect to the now outdated Thomistic version.

In pursuing this aim Buonafede divided the currents of 'modern' thought between the authors who deserved condemnation without reserve, whose ideas revealed an unbridgeable distance from the Christian philosophical-political outlook, and those who were 'correctable' and thus useful since they were not without some appreciable food for thought. In this operation Buonafede invoked his Augustinian conception, according to which a model of perfect virtue cannot exist on earth. As a result he sought to find the positive aspects of the authors he studied, which enabled him to 'save' some authors and works that other parts of the Catholic culture condemned without appeal, in particular through the Inquisition. For this reason Buonafede tended to condemn mainly the oldest writers— such as Machiavelli, Hobbes, Spinoza and Cocceji—while, besides Vattel, he 'saved' Montesquieu, Rousseau, Hume and Locke.[12]

Buonafede made only two direct references to Vattel in the *Conquiste celebri*, both in a chapter dedicated to Grotius that addressed the problem of the relationship between natural law and the law of nations. According to Buonafede, this problem and the potential incompatibility between the two concepts could only be resolved by using, as Vico had already done, the single expression "natural law of nations," which in fact he adopted in the title of his work. The natural law of nations could, in fact, make explicit the fact that the rules of behaviour for individual nations and individual people could never be detached from respect of the universal natural law, because they should recognise their foundation only in that. According to Buonafede, Grotious had "attenuated" this link by referring to a "positive law of nations" distinct from natural law and binding only through the principle of the *consensus gentium*. Grotius's law of nations thus became voluntary and "arbitrary." Precisely because of Vattel's distinction between a "necessary" and "voluntary" law of nations, however, Buonafede had no doubts about the fact that in the name of the unity and indivisibility of natural law even Grotius could continue to be used as an interpretative source of the right of nations, the right of war and the right of conquest.[13]

Buonafede's debt to Vattel was in fact much greater than what is expressly declared in the *Conquiste celebri*. Both in this text and in other works,[14] he shared Vattel's thesis on the legal and natural equality of nations within the interstate community, irrespective of their "power" (something which was obviously to the benefit of the "small" Papal State[15]). He agreed with the principle of the legitimacy of appropriating and making use of lands left abandoned and uncultivated by their "original" owners,[16] to the extent of accepting the idea of the right of resistance against a despot[17] and of the legitimacy of intervention, even military, against the "enemies of humanity."[18]

PROTESTANT ERRORS

Buonafede focused his interest on the *Droit des gens* twenty years later in another work dedicated to natural law and the law of nations. This was *Della restaurazione di ogni filosofia nei secoli XVI, XVII, e XVIII* (1785–1789),[19] which was then summarised in a more user-friendly book entitled *Della istoria critica del moderno diritto di natura e di genti* (1789).

These revealed the central principles of the theory of good government conceived by Buonafede with greater clarity, demonstrating that the author's main aim was to repudiate Hobbes's ideas on natural man as deprived of sensibility, which he deemed harmful to the natural harmony of society. In his opinion, Hobbes had confused sovereign power with absolute, arbitrary and unlimited power, such as to prejudice inalienable human rights and the laws of God.[20] This was a position that reflected the first ideas of an Enlightened socialism of the Italian space,[21] and represented a sort of third way, within Catholic culture, between an emphatic condemnation of the culture of the Enlightenment and the uncritical acceptance of Protestant natural law.

It is most significant that Buonafede showed renewed interest in Vattel during the 1780s. Loschi's Italian translation of the *Droit des gens* had been published in the intervening years, and we know that Buonafede and Loschi were close friends.[22] Indeed, there is an extensive correspondence between the two intellectuals preserved in the State Archives of Modena,[23] and Loschi had personally edited the reprints of some of Buonafede's works, including the *Conquiste celebri*. In effect, Buonafede and Loschi appear to have shared the overall cultural strategy that was behind the Italian translation of Vattel, and which more generally was linked to the desire to manage a careful and selective reception of Protestant natural law in Italy.

Buonafede also realised that the political and international context of the new readings and renewed success of Vattel had changed profoundly over the past thirty years. To his mind it made no sense to mount a defence of Catholicism by doing battle with the *Droit des gens*, accusing it of being rooted in Protestant theology. To be sure, he did not avoid pointing out certain passages of the *Droit des gens* which he believed to be a danger to the Catholic religion, but he immediately added that "the remainder," in other words the greater part of the work, was to be considered useful, well-documented and enjoyable.[24] For this reason he advised Italian readers to read Loschi's translation, with the added notes that he believed would give fair warning of incorrect interpretations. Buonafede also praised Vattel for his *Questions de droit naturel* and for his notes and comments on Wolff, appreciating the fact that he had helped to further moderate the German philosopher.[25] On the eve of the French Revolution, therefore, it seemed to Buonafede that the *Droit des gens* was still a text well worth consulting. However, with reference to the possibility of putting into practice all its principles and the fact that it did not always offer proof of the assertions it contained, he added that it should always be borne in mind that the work ought not to be taken completely seriously.[26]

THE RIGHT OF RESISTANCE AND THE RIGHT TO ASYLUM IN THE ITALIAN CONSTITUTIONS

In the years between the French Revolution and the birth of the Napoleonic Empire, the states of the Italian peninsula took part in an intense constitutional debate linked to the possibility of adopting written constitutions based on the model of the French ones, giving thought both to that of the constitutional monarchy of 1791 and the republican constitutions of 1793 and 1795. At the same time, the problem for the Italians was that of distinguishing themselves from France so as to preserve and promote the legal and philosophical tradition of the peninsula. Thus Vattel was once again at the centre of attention because the *Droit des gens* had been one of the sources and reference points for authors of the Italian Enlightenment (Cesare Beccaria, Gaetano Filangieri, Francesco Mario Pagano) and therefore served as a cultural bridge via which they could acquire the basic principles for their new constitutional texts.[27]

To Italian eyes the *Droit des gens* in fact offered a *trait d'union* between the Western natural law tradition and the republican conception of the constitutional state. It did so particularly in book one, where Vattel outlined a theory of the state and of constitutional sovereignty that had in those years been taken up in France by Sieyès. Even though it had not come into force, Sieyès's constitutional project was widely admired for the distinction it made between the various functions of the constituent power, which was responsible for determining the fundamental laws and functions of the constituted or legislative power that had to form the non-fundamental laws.[28] It is difficult to reconstruct with any precision all the genealogies through which the discourse on the natural rights of man moved, by way of the culture of natural law, to the written constitutions of the late eighteenth century.

Nonetheless, there is no doubt that the texts of the mid-century authors, including Burlamaqui and Vattel, played an important role within this consumption process of the natural law culture. Not by chance, as other authors have pointed out it was Burlamaqui who coined the expression "Droits de l'homme" (1747), and hence the English "rights of man," which came from the London translation of 1748.[29] Vattel—like most of his contemporaries—did not yet use the expression, but the *Droit des gens* contained a catalogue of those rights and the rights of the citizen, which reinterpreted the natural law tradition, and which his contemporaries had no difficulty in interpreting in combination, together with the work of Burlamaqui and other authors.

The debate that took place in the Italian republics after Napoleon's army had entered Milan (1796) revolved around two crucial elements: the first was that of how to formulate the *Dichiarazioni dei diritti* so as to avoid them being a simple copy of the French declaration of 1789. This concern explains why the declarations became statements about the rights *and duties* of man and citizen, thus exhibiting the influence of natural law. The other crucial element related to the list of rights and the choice of which to include in or omit from the declarations and the constitutions. And it is in this debate that the contribution and practical use of the *Droit des gens* can be most accurately identified.

This reasoning applies in particular to the inclusion in the constitutions of the right of resistance and the right to asylum. The key text in the Italian and Mediterranean sphere was the *Progetto di costituzione per la repubblica napoletana presentato al governo provvisorio nel 1799*, which was important because it was made up of three parts: the report from the committee charged with drafting it, the declaration of rights and duties, and the constitution. The text was written in large part by Francesco Mario Pagano, the most important of the friends and a pupil of Gaetano Filangieri. Unfortunately the Neapolitan constitution never came into force because the republic was defeated by the Bourbons and their English allies and its protagonists were all hanged, but it is nevertheless possible to identify within it the legacy of the culture of the Italian Enlightenment and the influence of the natural law tradition handed down by Vattel.

The first point to which we must pay attention concerns the formulation of the right of resistance. Francesco Mario Pagano's declaration of human rights is the only Italian text to define this as a right of man, since in general the other Italian declarations, modelled on the French constitution of the Year III, only contained a definition of the arbitrary acts committed in violation of the rights of man, namely those carried out by the political authority outside of the examples and conventions established by the law. It is true that the Declaration of Independence of the United States of America had previously stated that one of the natural and imprescriptible human rights was the freedom to resist oppression, but the Neapolitan text went into greater detail, seeking to distinguish between man in the state of nature and man in the social state. According to Pagano, therefore, the natural right of resistance as a human right had to remain inactive as long as the man and the citizen acted independently, since otherwise the result would have been to encourage anarchy. The right of resistance was activated instead in the political sphere, when what was a natural law became a civil right and was exercised collectively by the entire population against any tyrannical authority that abused its constitutional powers.[30] Accordingly, in the Neapolitan declaration we find a typical formulation of the right of resistance that derives not so much from the revolutionary culture as from the natural law tradition, in the sense that it is not an expression and the consequence of the popular will, but, as it was for Vattel, the reflection of a natural right resulting from the abuse of power and the excess of tyranny that end up violating constitutional guarantees.[31]

The other point is that of whether the right to asylum was a human right, and through this we can again measure the practical use of the *Droit des gens* in the constitutional practices of late eighteenth-century Italian republicanism. The right to asylum and the safeguarding of foreigners was a highly controversial subject in the legal culture of Mediterranean countries. Traditionally, the granting of asylum was considered a means of asserting political power, given that it was born as a religious institution, in the form of a power and thus of a concession dependent on the sovereignty and authority of the Catholic Church, first in the exercise of religious power and then of temporal power. The culture of European natural law had then devoted much attention to the right to asylum construed as a protection of foreigners, reflecting on its compatibility with the full exercise of sovereignty by a state and affirming the existence of a right of access for foreigners seeking refuge in the territory of another state on the basis of natural law and the *ius gentium*. Grotius had thus clarified the dual nature of asylum, as a natural law of the subject and as a prerogative of the sovereign in granting it. Pufendorf had studied the matter in depth, recognising the solidarity and the humanitarian character of asylum, but without recognising the existence of a natural law obligation for the state to receive and grant shelter indiscriminately to all foreigners. On the one hand, therefore, hospitality became a common human duty, while on the other the interest of the host state was safeguarded for the sake of its particular interests. Wolff in turn had confirmed the imperfect nature of this right due to the fact that it was subordinated to the discretionary judgement of the sovereign.[32]

In the *Droit des gens* this picture was complicated further since Vattel linked the right to asylum to a series of other freedoms and rights that measured itself against political authority. Asylum became a point of intersection between the problem of detachment from one's homeland (voluntary, imposed or ordered) and that of arriving in other countries (expressed through asylum, but also through welcome and shelter).[33] The request for asylum could be the consequence of a judicial sentence (such as banishment and forced exile), or it could be a right of a person who left his country not as a result of a conviction, but as the voluntary exercise of a specific personal right that could be considered a form of passive resistance. From this derived, in the *Droit des gens*, the analysis of the duties of nations towards those who required asylum, and of the right to determine whether or not to accept a foreigner in one's territory. According to Vattel, a nation could not refuse, in the name of natural law, the even perpetual acceptance of a person expelled from his homeland, and if that person was denied entry it had to be for precise reasons that were carefully examined: among these could be the impossibility of being able to provide for the needs of both the nation and the foreigner at the same time.

While Vattel, as we can see, attempted to attach the recognition of the right to asylum to natural laws, some of his contemporaries lined up on the opposite side. First among these was Cesare Beccaria, who knew the *Droit des gens* well but openly argued in favour of the abolition of asylum as a manifestation of sovereignty, because he considered it to be both a danger to social order and an obstacle to the full administration of justice, even though he did show some sympathy to the welcoming of foreigners fleeing from tyrants.

The practice of the right to asylum and its translation into a human right was therefore forcefully discussed and combatted by opposing needs. From 1769, most of the European states, and especially the Italian ones, ended up abolishing the right to asylum in the name of the privileges of sovereignty, the sole exception being the Papal States. Nevertheless, the debate remained open and a section of the Italian Enlightenment culture continued to emphasise asylum as an inviolable human right, as, for instance, Francesco Mario Pagano did in one of his *Saggi politici* (1791), borrowing some ideas from Gaetano Filangieri.[34]

Within the constitutional tradition and that of international law, the *Droit des gens*, in which the right to asylum was presented and discussed, continued to serve as an important cultural bridge, thereby facilitating the inclusion of this right in the general category of human rights.[35] A demonstration of this, once again, was the draft Neapolitan constitution of 1799, in which the exercise of the right to asylum was reintroduced (albeit confined to the general dispositions), in the wake of the provisions of the draft French constitution of 1793. Vattel's reflection on asylum would continue to be at the root of Italian political thought through the nineteenth century, until it merged into the preparatory work for Article 10 of the Republican Constitution of 1948, which definitively sanctioned the protection of the right to asylum as a right of the foreigner on Italian soil.[36]

NOTES

1. Lynn Hunt, *Inventing Human Rights: A History* (New York: Norton & Co., 2008).
2. John Robertson, *The Case for Enlightenment: Scotland and Naples 1680–1760* (Cambridge: Cambridge University Press, 2007), 386–400; Vincenzo Ferrone, *The Politics of Enlightenment: Constitutionalism, Republicanism and the Rights of Men* (London-New York: Anthem Press, 2012), 197–219.
3. Antonio Trampus, *"Enlightenment in Global History: On Filangieri's Science of Legislation and the Transformation of Political Language in the Classical Liberalism,"* in Век Просвещения. Что такое Просвещение? Новые ответы на старый вопрос (Le Siècle des Lumières. Qu'est-ce que les Lumières? Nouvelles reponses à l'ancienne question), ed. Serguei Karp *et al.* (Moscow: Nauka, 2019), 110–125.
4. Davis Hume, "Of the Jealousy of Trade," in *Essays and Treatises on Several Subjects, a new edition* (London: Millar, 1758) and Istvan Hont, *Jealousy of Trade: International Competition and the Nation-State in Historical*

Perspective (Cambridge, MA: Harvard University Press, 2005); Gaetano, Filangieri, *La scienza della legislazione. Edizione critica,* 7 vols., dir. Vincenzo Ferrone, eds. Antonio Trampus *et al.* (Venice: Edizioni della Laguna 2004), book 1, chap. XX.

5. Vincenzo Ferrone, *Storia dei diritti dell'uomo. L'Illuminismo e la costruzione del linguaggio poitico dei moderni* (Rome-Bari: Laterza, 2014), 330.

6. Ferrone, *The Politics of Enlightenment,* 47, 234.

7. Filangieri, *La scienza della legislazione,* book 4, ch. 43.

8. Ulrich L. Lehner, *The Catholic Enlightenment: The Forgotten History of a Global Movement* (Oxford: Oxford University Press, 2016).

9. Ilario Tolomio, *Theism and the History of Philosophy: Appiano Buonafede,* in *Models of the History of Philosophy, III. The Second Enlightenment and the Kantian Age* (New York: Springer, 2015), 359–382.

10. Giulia Delogu, "The Political Functions of Virtue in the Eighteenth-Century Italian Debate," *History of European Ideas* 43 (2017), 889–913.

11. Alberto Clerici, "Vattel in the Papal States: The Law of Nations and Anti-Prussian Propaganda in Italy at the Time of Seven Years' War," in *The Legacy of Vattel's Droit des gens,* ed. Koen Stapelbroek and Antonio Trampus, 207–234.

12. Appiano Buonafede, *Delle conquiste celebri esaminate col diritto natural delle genti* (Lucca: Riccomini, 1763), 18, 25, 78; On the Augustinian reading of Buonafede see Giulia Delogu, *La poetica della virtù. Comunicazione politica e rappresentazione del potere in Italia tra Sette e Ottocento* (Milan: Misesis, 2017), 122.

13. Buonafede, *Conquiste celebri,* 65.

14. Clerici, "Vattel in the Papal States," 219–222.

15. Buonafede, *Conquiste celebri,* 42.

16. Buonafede, *Conquiste celebri,* 229–230, with reference to Pufendorf and Vattel, *Droit des gens,* book I, ch. VII, § 81 and book II, ch. VII, § 86–87, 97.

17. Buonafede, *Conquiste celebri,* 205–215; Vattel, *Droit des gens,* book I, ch. IV, § 51–54.

18. Buonafede, *Conquiste celebri,* 139 and Vattel, *Droit des gens,* book III, ch. III § 34 and book IV, ch. I, § 5.

19. The German translation of the three volumes of the *Restaurazione* was published in Leipzig by a Kant scholar, Karl Heinrich Heydenreich. On this translation see also Sophus A. Reinert, *The Academy of Fisticuffs: Political Economy and Commercial Society in Enlightenment Italy* (Cambridge Mass.: Harvard University Press, 2019), 295–98.

20. Cromaziano Agatopisto [Appiano Buonafede], *Della restaurazione di ogni filosofia nei secoli 16., 17., 18.,* vol. 2 (Naples: Porcelli, 1788), 51.

21. Reinert, *The Academy of Fisticuffs,* 296–97.

22. Agatopisto [Buonafede], *Della restaurazione di ogni filosofia*, 222.
23. Antonio Trampus, "Il ruolo del traduttore nel tardo Illuminismo: Lodovico Antonio Loschi e la versione italiana del "Droit des gens" di Emer de Vattel," in *Il linguaggio del tardo Illuminismo. Politica, diritto e società civile*, ed. Antonio Trampus (Rome: Edizioni di Storia e Letteratura, 2011) 81–108. The letters are in ASM, Archivio per materie. Letterati-Lodovico Antonio Loschi.
24. Agatopisto [Buonafede], *Della restaurazione di ogni filosofia*, 223.
25. Emer de Vattel, *Questions de droit naturel, et observations sur le Traité du droit de la nature de M. le baron de Wolf, par M. de Vattel* (Bern: Société typographique, 1762).
26. Agatopisto [Buonafede], *Della restaurazione di ogni filosofia*, 223.
27. For the systematic use of Vattel by Beccaria see *Des délits et des peins* ed., Philippe Audegean (Lyon: ENS, 2009) 312–13, 317, 323, 330, 333, 352, 356, 361, 368, 377, 384, 388, 393.
28. Oliviere Beaud, *La Puissance de l'État* (Paris: Presses Universitaires de France, 1994), 206.
29. Hunt, *Inventing Human Rights*, 231.
30. *Progetto di costituzione della Repubblica napoletana presentato al governo provvisorio dal comitato di legislazione*, Federica Morelli and Antonio Trampus, eds., introduction by Anna Maria Rao (Venice: Centro di Studi sull'Illuminismo europeo, 2008), 66, 123–24, 132.
31. Vincent Chetail, "Vattel et la sémantique du droit des gens: une tentative de reconstruction critique," in *Vattel's International Law in a XXIst Century Perspective / Le droit International de Vattel vu du XXIe siècle*, ed. Vincent Chetail and Peter Haggenmacher (The Hague: Martinus Nijhoff, 2011), 405.
32. Ludwig von Bar, *A History of Continental Criminal Law* (Boston: Little, 1916), 88–92; Gil Loescher, ed., *Refugees and the Asylum Dilemma in the West* (Pennsylvania: Pennsylvania State University Press, 1992); Fabrizio Mastromartino, *Il diritto d'asilo. Teoria e storia di un istituto giuridico controverso* (Turin: Giappichelli, 2012).
33. Alberto Carrera, "The citizen's right to leave his country: The concept of exile in Vattel's *Droit des gens*" in *The Legacy of Vattel's Droit des gens*, ed. Koen Stapelbroek and Antonio Trampus (Basingstoke: Palgrave Macmillan, 2019).
34. Francesco Mario Pagano, *Saggi politici. De' principii, progressi e decadenza della società (1791–92)*, ed. Luigi Firpo and Laura Salvetti Firpo (Naples: Vivarium, 1993), 262 (essay IV, chap. II).
35. Manuel R. García Mora, *International Law and Asylum as a Human Right* (Washington: Public Affairs Press, 1956), 38–40; Peter Paul Remec, *The Position of the Individual in International Law according to Grotius and*

Vattel (The Hague: Martinus Nijhoff, 1960), 293; Atle Grahl-Madsen, *The Status of Refugees in International Law: Asylum, entry and sojourn* (Leiden: Sijthoff, 1966), vol. 2, 16.

36. Pietro Costa, *Costituzione italiana: articolo 10* (Rome: Carocci, 2018). For a comprehensive discussion on this right with reference to Vattel see Luke Glanville, "Historical Thinking about Human Protection. Insights from Vattel," in *Routledge Handbook of Ethics and International Relations*, ed. Brent J. Steele and Eric A. Heinze (London: Routledge, 2018), 308–317.

BIBLIOGRAPHY

Most of the quotations of Vattel's *Droit des gens* come from the widely available English-language edition of 2008, which includes an introduction by Béla Kapossy and Richard Whatmore (Indianapolis: Liberty Fund) and maintains the text and English title *The Law of Nations: Or, Principles of the Law of Nature Applied to the Conduct and Affairs of Nations and Sovereigns* of the 1797 London standard edition.

Contemporary translations of primary sources are listed under the names of their authors. Full manuscript sources are referenced only in the notes for reasons of space.

ARCHIVAL ABBREVIATIONS

ASM Archivio di Stato di Modena, Modena, Italy

PRIMARY SOURCES

PRINTED BOOKS

Beccaria, Cesare, *Des délits et des peines*, ed. Philippe Audegean (Lyon: ENS, 2009)

Buonafede, Appiano, *Della restaurazione di ogni filosofia nei secoli 16., 17., 18.*, vol. 2 (Naples: Porcelli, 1788)

———, *Delle conquiste celebri esaminate col diritto natural delle genti* (Lucca: Riccomini, 1763)

Filangieri, Gaetano, *La scienza della legislazione. Edizione critica*, 7 vols., dir. Vincenzo Ferrone, eds. Antonio Trampus *et al.* (Venice: Edizioni della Laguna 2004)

Hume, David, *Of the Jealousy of Trade*, in *Essays and Treatises on Several Subjects, a new edition* (London: Millar, 1758)

Pagano, Francesco Mario, *Saggi politici. De' principii, progressi e decadenza della società (1791–92)*, eds. Luigi Firpo and Laura Salvetti Firpo (Naples: Vivarium, 1993)

Progetto di costituzione della Repubblica napoletana presentato al governo provvisorio dal comitato di legislazione, eds. Federica Morelli and Antonio Trampus, introduction by Anna Maria Rao (Venice: Centro di Studi sull'Illuminismo europeo, 2008)

Vattel, Emer de, *Law of Nations: Or, Principles of the Law of Nature, Applied to The Conduct and Affairs of Nations and Sovereigns*, eds. Béla Kapossy and Richard Whatmore (Indianapolis: Liberty Fund 2008)

———, *Questions de droit naturel, et observations sur le Traité du droit de la nature de M. le baron de Wolf, par M. de Vattel* (Bern: Société typographique, 1762)

SECONDARY SOURCES

Bar, Ludwig von, *A History of Continental Criminal Law* (Boston: Little, 1916)

Beaud, Oliviere, *La Puissance de l'État* (Paris: Presses Universitaires de France, 1994)

Carrera, Alberto, "The citizen's right to leave his country: The concept of exile in Vattel's *Droit des gens*" in *The Legacy of Vattel's Droit des gens*, ed. Koen Stapelbroek and Antonio Trampus (Cham: Palgrave Macmillan, 2019), 77–93

Chetail, Vincent, "Vattel et la sémantique du droit des gens: une tentative de reconstruction critique," in *Vattel's International Law in a XXIst Century Perspective / Le droit International de Vattel vu du XXIe siècle*, eds. Vincent Chetail and Peter Haggenmacher (The Hague: Martinus Nijhoff, 2011), 385-434

Clerici, Alberto, "Vattel in the Papal States: The Law of Nations and Anti-Prussian Propaganda in Italy at the Time of Seven Years' War," in *The Legacy of Vattel's Droit des gens*, ed. Koen Stapelbroek and Antonio Trampus (Cham: Palgrave Macmillan, 2019) 207–234

Costa, Pietro, *Costituzione italiana: articolo 10* (Rome: Carocci, 2018)

Delogu, Giulia, "The Political Functions of Virtue in the Eighteenth-Century Italian Debate," *History of European Ideas* 43 (2017), 889–913

———, *La poetica della virtù. Comunicazione politica e rappresentazione del potere in Italia tra Sette e Ottocento* (Milan: Misesis, 2017b)

Ferrone, Vincenzo, *Storia dei diritti dell'uomo. L'Illuminismo e la costruzione del linguaggio poitico dei moderni* (Rome-Bari: Laterza, 2014)

———, *The Politics of Enlightenment: Constitutionalism, Republicanism and the Rights of Men* (London-New York: Anthem Press, 2012)

García Mora, Manuel R., *International Law and Asylum as a Human Right* (Washington: Public Affairs Press, 1956)

Glanville, Luke, "Historical Thinking about Human Protection. Insights from Vattel," in *Routledge Handbook of Ethics and International Relations*, eds. Brent J. Steele and Eric A. Heinze (London: Routledge, 2018) 308–317

Grahl-Madsen, Atle, *The Status of Refugees in International Law: Asylum, entry and sojourn* 2 vols. (Leiden: Sijthoff, 1966)

Hont, Istvan, *Jealousy of Trade: International Competition and the Nation-State in Historical Perspective* (Cambridge, Mass.: Harvard University Press, 2005)

Hunt, Lynn, *Inventing Human Rights: A History* (New York: Norton & Co., 2008)

Lehner, Ulrich L., *The Catholic Enlightenment: The Forgotten History of a Global Movement* (Oxford: Oxford University Press, 2016)

Loescher, Gil, ed., *Refugees and the Asylum Dilemma in the West* (Pennsylvania: Pennsylvania State University Press, 1992)

Mastromartino, Fabrizio, *Il diritto d'asilo. Teoria e storia di un istituto giuridico controverso* (Turin: Giappichelli, 2012)

Reinert, Sophus A., *The Academy of Fisticuffs: Political Economy and Commercial Society in Enlightenment Italy* (Cambridge, Mass.: Harvard University Press, 2019)

Remec, Peter Paul, *The Position of the Individual in International Law according to Grotius and Vattel* (The Hague: Martinus Nijhoff, 1960)

Robertson, John, *The Case for Enlightenment: Scotland and Naples 1680–1760* (Cambridge: Cambridge University Press, 2007)

Tolomio, Ilario, *Theism and the History of Philosophy: Appiano Buonafede*, in *Models of the History of Philosophy, III. The Second Enlightenment and the Kantian Age* (New York: Springer, 2015), 359–382

Trampus, Antonio, "Il ruolo del traduttore nel tardo Illuminismo: Ludovico Antonio Loschi e la traduzione italiana del *Droit des gens*," in *Il linguaggio del tardo Illuminismo*, ed. Antonio Trampus (Rome: Edizioni di Storia e Letteratura, 2009)

———, *"Enlightenment in Global History: On Filangieri's Science of Legislation and the Transformation of Political Language in the Classical Liberalism"*, in Век Просвещения. Что такое Просвещение? Новые ответы на старый вопрос (*Le Siècle des Lumières. Qu'est-ce que les Lumières? Nouvelles reponses à l'ancienne question*), ed. Serguei Karp et al. (Moscow: Nauka, 2019), 110–125

CHAPTER 10

Bern, the French Revolution and the Congress of Vienna

Another unpublished source particularly helpful for understanding the reasons behind appropriating Vattel's work in seventeenth- and eighteenth-century Europe is one that initially appears to refer to Vattel's own Switzerland. This is the French-language manuscript housed in Bern's Bürgerbibliothek, which is in fact a transcription of the *Droit des gens* presumably of around 1780 and also presumably based on the Neuchâtel edition of 1773.[1] The setting of the Swiss confederation is of relevance not only because it was Vattel's homeland, but also because it was a small state established as a republic that in the course of the eighteenth century, shortly after the death of the author, sought to reposition itself in the international system and in the politics of neutrality by inspiring its intellectuals and organising the ruling classes in defence of ancient freedoms.[2]

The Bern Manuscript: Re-Contextualising Small States

The uses and interpretations made of the *Droit des gens* in this context are therefore unsurprising, even though it has yet to undergo specific study and many relevant sources still need to be investigated.[3] One of these, as

Archival Abbreviations
ASN   Archivio di Stato di Napoli, Naples, Italy
BB   Bürgerbibliothek Bern, Berne, Switzerland

© The Author(s) 2020   173
A. Trampus, *Emer de Vattel and the Politics of Good Government*,
https://doi.org/10.1007/978-3-030-48024-0_10

we shall see, raises certain particularly interesting and not entirely resolved questions about the actual context of its production, and it is precisely the manuscript volume held in the Bernese library, which bears a shortened version of the title, *Droit des gens par Mr de Vattel*, but gives no indication of the compiler, or any introductory note or hint that might help identify the edition from which it was drawn.[4] On the back of the hardbound front cover there is a late eighteenth-century or early nineteenth-century aristocratic *ex libris* containing the initials "A.T." The catalogue of the Bürgerbibliothek traces the document to a bequest by Johann Anton von Tillier, suggesting that it was used as study material. The library also preserves other manuscripts once belonging to the same Tillier, including an account of an educational journey taken in 1785 to Hamburg and Brunswick.[5] Tillier (1760–1810) belonged to a noble family of Bern[6] and at the time of his trip was a student in Gottingue.[7] After returning home, presumably having graduated, he worked for the Bern administration, later becoming a *ratsherr* (alderman) in the city.[8] In 1787 he married Anna Elisabeth Tscharner (1768–1843), who also belonged to a Bernese noble family. The attribution of the manuscript to Johann Anton von Tillier, in the absence of other clues, is therefore based on its being located in the Bernese archive and the identifying record of the archivists. However, it is probably also worth noting in passing that the Tillier family appears to have had direct contact with Vattel, as testified in the *Mémoires politiques & militaires pour servir à l'histoire de notre tems*, a text often attributed to Vattel.[9]

The manuscript attracts attention because it poses problems of interpretation of a particular kind. The first regards the possibility of identifying the edition used by the copier because between the publication of the *Droit des gens* in 1758 and the mid-1770s, which as we shall see was presumably when the manuscript was produced, there appeared many editions of the work quite unlike each other.[10] Their differences were not only insignificant variations in the text, but mainly concerned the typographical composition in the distribution of the text and in the way the chapters and sections were numbered. A deeper examination of the manuscript based on its typographical organisation reveals that its structure and textual organisation relied directly on the Neuchâtel edition of 1773.

The manuscript, however, is not a simple transcription of the printed text: first of all it does not include the *Préface* of around thirty pages, nor does it contain the full *Table des matières*, which is placed in simplified form at the end of the text instead of immediately after the *Préface*, as in the original. There is also no trace of the *Abregé de la vie de M. de Vattel* that appears at the start of the second printed volume.[11] The manuscript also lacks the titles of the sections, which were placed in the margins of the printed version, but it retains their numbers.

The most striking characteristic of the document is, however, the fact that it is not a complete transcription of the text, but only a sort of abstract, in which only the first few sentences of each section are reproduced along with, in the case of the particularly long passages, the sentences considered most important. These cuts and selections reduce the overall length of the work to around half of the original, thus producing a kind of abridgement, apparently intended for some kind of practical use or for study. We will return to consider these features further on, but at this point it is worth observing that all this, in addition to the fact that the text is included in a collection of manuscripts belonging to Tillier and dating back to his student years in Gottingue, seems to suggest that it was indeed created for the purpose of academic study. Moreover, the fact that in that period a student like Tillier, who because of his Swiss origins would have had no difficulty in obtaining Swiss editions of the *Droit des gens*, used the Neuchâtel one of 1773 offers more interesting evidence about the circulation of the work.[12]

A second interesting characteristic of the Bern manuscript is the fact that the transcriber not only made selections of the text, but also inserted within it new information and examples not present in the original. These were taken mostly from events of or near his own time and were aimed at contextualising or re-contextualising the general principles propounded by Vattel. When adapting the text for his university studies, Tillier—if he was indeed the compiler of the abregé—did exactly what Vattel said, in his letters to Malesherbes in 1757, he feared the pirate publishers would do (that is "slip in reflections or examples that I would not want to admit": "d'y glisser des réflexions ou des examples, que je ne voudrois pas avouer"),[13] running the risk that they would bend the interpretation of *Droit des gens* to support ongoing conflicts or international problems.

Tillier's manuscript, where it abbreviated Vattel, adhered to the original. The copying is faithful, and even when the transcriber summarises he does so using Vattel's own words, with only rare or insignificant instances of linguistic licence. Even the notes match the original, sometimes being incorporated in parentheses or sometimes, more rarely, being written in the margins. It is in fact one of these notes, which refers to the issue of the ambassador's exemption from the civil jurisdiction of the country in which he resides (making reference to an actual incident in England in 1658 to the detriment of the ambassador from Brandenburg), that reveals that the edition of the *Droit des gens* used for Tillier's synthesis was in fact that of 1773. This is because the note cites as a source, alongside the *Traité du juge* by Cornelis van Bynckershoek, the 1771 annal of Bouillon's *Journal politique*, which is absent from the first edition of 1758 and was first inserted in the 1773 Neuchâtel edition.[14]

The context in which the Bern manuscript was produced is particularly complex. On the one hand, the document reflects cultural circumstances of the Swiss environment from which Tillier came, but it is not this that must be investigated in order to discern the contingent reasons that prompted the abridgement.[15] Rather, attention should focus on the university environment of Gottingue, which gives rise to another problem in the study of the spread and reception of Vattel's work, namely that of its circulation and use not in the complete and original form desired by the author, but in summaries, abstracts and extracts, especially in university settings intended for the education of the ruling classes of the Napoleonic age and the Restoration. This is an avenue of research mostly unexplored, for the attention of scholars has been mainly oriented towards the development of the science of the state.[16] Even so, it is an area that merits more detailed study in order to grasp the actual influence of the *Droit des gens* on the political culture of the early nineteenth century.[17] In the case of Italy, for example, the persistent use of Vattel's book in the teaching of the law of nations explains the 1804 reprint that appeared in Bologna—another context linked to the tradition of the small state—of the translation attributed to Lodovico Antonio Loschi,[18] as well as the spread of the text in Bolognese academia and the decision of Giuseppe Prina, professor of the law of nations at the University of Pavia from 1804 to 1818, to use Vattel given that he deemed him the most brilliant pupil of Christian Wolff and the founder of international public law.[19]

VATTEL THE REVOLUTIONARY

In effect, the consequences of the French revolution led Italians to interpret the *Droit des gens* in a new and revolutionary way, in the wake of the observations, for example, of Verri, who—as we have seen—recognised in it an authoritative justification for democratic practices and for affirming the right of a people to choose its form of government.

Another sign of Vattel's presence in Italy during the closing years of the eighteenth century can be found in the *Sistema universale dei principi del diritto marittimo dell'Europa* by the Sardinian lawyer Domenico Alberto Azuni, who later collaborated with Napoleon and was in charge of drafting the maritime and commercial code of 1808. Azuni had already come across Vattel's work in his *Dizionario universale ragionato della giurisprudenza mercantile* (1786–1788), concurring with his idea that there was a voluntary natural law alongside the immutable law that derived from nature itself. He explained that in wartime neutral states had every right to continue trading with warring powers, because this was consistent with the principle of independence and equality among all nations. Only the voluntary law of nations, arising from a convention or specific treaties, could bring about the opposite.[20]

In the second volume of the *Sistema universale*, published in Trieste in 1797, Azuni first criticised the weak points of Vattel's work, especially with regard to neutrality, as his position seemed to cling too closely to the principles of Grotius, Gronovius, Barbeyrac, Coccejus and Heineccius, resulting in an excessive prudence, confusion and lack of clarity in the chain of events.[21] It is evident that up to that point Azuni had simply reiterated the observations made fifteen years earlier by Ferdinando Galiani. However, later in the text he expressed his appreciation of the contribution of Vattel's work to the debate on neutrality, saying that thanks to the *Droit des gens* it had been possible to arrive at a clear definition of two types of neutrality: active neutrality, which occurs when the neutral state pursues trade in order to profit from a state of war between two nations; and passive or perfect neutrality, adopted in the past by many European states. As Azuni explained, Vattel's work had by then received the ultimate accolade fully ten years after the death of its author. It was in fact the very model of neutrality described in the *Droit des gens* which was imposed in the new international politics of the latter half of the eighteenth century, and which was adopted with the manifesto of 1 August 1778 by Tuscany in the war between Great Britain and the American colonies, France, Holland and Spain, by Venice with the declaration of neutrality of 9 September 1779, by Naples with the royal edict of 19 September 1778, by Genoa with the declaration of 1 July 1779, and by the Papal States with the edict of 4 March 1779.[22]

Almost at the same time as Verri's reflections, in Naples Bernardo Tanucci, minister of finance in the Kingdom of Naples, had already grasped that the use of Vattel could be a danger to the orderly administration of justice and "good government" in the Italian states.[23] Indeed, commenting on a decision made by the president of the supreme court of justice in Florence, which had based its arguments on Genovesi, Vattel, Montesquieu and Lampredi, Tanucci had observed that these were "modern people," whose ideas and theories had not yet been sufficiently tested and which did not enjoy the recognition of much of European culture and thus lacked authority.[24]

As a result of the events of the French Revolution Vattel's work was read in Naples in an increasingly critical manner. In 1790 the abbot Michelangelo Grisolia (1751–1794), a professor of ethics in the Naples military academy and author of texts in favour of the absolute power of the Neapolitan monarchy, had presented lessons on public law in which he explicitly challenged the use of the term 'nation' in the *Droit des gens* since it seemed to him to be irredeemably in conflict with the concept of state to which the Kingdom of Naples itself aspired.[25] He devoted considerable time, in his early lessons, to the *Droit des gens*, concentrating almost exclusively on the problem of the nation and the conflicting relationship between the nation and the state of the prince. According to Grisolia, Vattel's main mistake was his failure to realise that the principle of the nation came into conflict with the idea of a patrimonial state ("regno patrimoniale"), in which the nation itself belonged to the prince and sovereignty was inalienable, with the result that the nation unavoidably found itself subordinate to the state and the sovereign. For Grisolia, the time had come to call into question Vattel's authority, because "time greatly influences the possibility of courageously supporting certain truths or, on the contrary, rejecting them."

A few months later, in Naples the government censor Giovanni Francesco Conforti—who later, however, adhered to revolutionary ideals—suggested a ban on the distribution of the *Droit des gens* because it was an "incendiary" work that spread the same ideas that had led the "insane" philosophers to rebel against the hereditary monarchy. In his notes of December 1792 on the books he was reading and censoring, Conforti still showed full confidence in the monarchy, surely reinforced partly by fear of what was happening in Paris. This was in fact the moment in which Louis XVI and Marie Antoinette, sister of Queen Maria Carolina of Naples, had just been deposed and the trial that led the sovereign to the guillotine was being prepared. Vattel had to be condemned, Conforti said, because he proposed a completely inaccurate interpretation of the theory of popular sovereignty and of the recognition of the rights of the "whole

body of the people." His considerations seemed to be almost a justification of the government of Louis XVI and consequently also of the monarchy of the Kingdom of Naples. In fact, in Conforti's opinion, according to
the culture of natural law and the public law of Grotius and Pufendorf, in
hereditary monarchy the nation had no right to influence the constitution
and government because only the monarch was ordained by God to maintain order and ensure the happiness of the people.[26]

These observations are interesting because by excluding Vattel from the
tradition of classical natural law, they clearly demonstrate that a republican
interpretation of the *Droit des gens* had taken root. It is important to
remember that, in the wake of the Revolution and the fall of the monarchy, Vattel had recovered his relevance on both sides of the Atlantic,
because he offered useful arguments to the Americans and to the French
against the aggressive politics of Great Britain, and because he justified
recourse to war as an essential act that complemented the exercise of
sovereignty.[27]

Another interesting reading of Vattel's work in the aftermath of the
Revolution is found in Bologna, the Italian city which until 1796 had been
a legation of the Papal States, which is to say a sort of papal enclave located
in the Po Valley. As a small city-sized state, subject to the Pope but governed by a citizen patriciate in accordance with the dynamics of mixed
government, Bologna had over the centuries developed a political tradition based on the relationship between the exercise of monocratic government and the need for depositories of laws to exist within a monarchy (like
that of the Pope) of intermediate political bodies (like cities). Thus in the
eighteenth century Montesquieu's theories had been favourably received,
including even those relating to the political function of the nobility conceived as a bulwark against a despotic government, and over time
Bolognese culture had given significant attention to the debate on the
right of resistance.[28] The Geneva Revolution of 1782 had been followed
with particular attention and the European newspapers, among them the
Journal des gens du monde in Frankfurt, had commented positively on the
case of Bologna where the ability of the citizen government to maintain a
balance between the rights of the city and those of the government of the
Holy See conveyed anti-tyrannical feelings in a constructive way.[29]

The French occupation of Bologna in 1796 was hailed as an opportunity to entrust the city to the protection of a republic, the French one—
which had numbered among the natural and inalienable rights of man the
right to resist oppression—in part because the city had long felt itself to

be, although subject to the Pope, a kind of republic. Vattel's *Droit des gens* was relevant to these arguments, especially because of the part in which it attached the characteristics of freedom and sovereignty to the existence of a constitutional state endowed with fundamental laws that underpinned the nation and the freedom of men, in the sense that already using the word 'nation' meant using the word 'freedom.'[30]

It is therefore understandable that Loschi's translation of Vattel's *Droit des gens* was reprinted in Bologna in 1804. Twenty years had elapsed since the Venetian edition, during which time Loschi had completed his entire political career, and then, after returning to nearby Modena, had become a member of the committee for the constitution of the Cisalpine Republic (1797) and provisional president of the Municipality of Modena (1800). The re-edition of Vattel's work was not, however, a mere act of homage to his standing as a man of letters, translator and politician, but rather indicated a new cultural strategy of appropriation of the text, decontextualised with respect to the Venice of the 1780s and re-contextualised in the republican Bologna. The new edition took the form of a textbook, with a format suited to study rather than in the small, almost pocket, size that in Venice had made it look more like a literary and political work. Loschi's name remained in the frontispiece, but all the dedications of the Venetian edition that had placed Vattel within the cultural framework of late eighteenth-century republicanism had been removed, along with the translator's notes. It seems that the edition was aimed at the students of the city's university, since the printers were the Masi brothers, with their bookshop "under the porticoes of the schools." An interesting point about the publisher is that the printing press of the brothers Glauco and Tommaso Masi was that which up to the end of 1800 had also been active in Livorno where it published, under the false place-name of Philadelphia, many important texts of the Italian and European Enlightenment, including Filangieri's *Scienza della legislazione*.[31]

The genesis of the 1804 Italian edition of Vattel's work was therefore certainly linked to the transformations taking place in the city that had belonged to the Pope, but it also belonged in a broader context of the crisis of revolutionary republicanism that foreshadowed the transformation of France into an empire and Italy into a kingdom dependent on it. In fact, in January 1802, the Italian Republic with Napoleon as provisional president was born, but already during 1804 its transformation into the Kingdom of Italy, formalised by the coronation of Napoleon in Milan in May 1805, was looming.

Another link between Vattel's book and Filangieri's emerged at the start of the nineteenth century through the work of Donato Tommasi, another old pupil of the Neapolitan philosopher, as well as a Mason and member of the order of the Illuminati. Tommasi had spent a career in the administration of the Kingdom of Naples, and when the Bourbons were forced to move to Sicily due to the French occupation of Naples he became a trusted figure for the royal family, then a minister in the royal house and finally the minister for the interior. In 1808 he accompanied in his role as tutor prince Leopold of Bourbon, son of King Ferdinand IV of Bourbon and of Maria Carolina of Habsburg, on a journey to Spain and Gibraltar, in an attempt to bring the Spanish throne to the Bourbons of Naples. During the journey, as Tommasi himself wrote to Queen Maria Carolina, he taught the young prince Leopold the fundamental characteristics of a monarchical institution through a reading of Vattel and extracts from other works, probably including those by Filangieri.[32]

Vattel's work was destined to play another important role in the life of Bologna after the fall of Napoleon. With the Congress of Vienna, the city was once more governed by the Pope, who in 1816 made arrangements for a reorganisation of the government of the city and its territory that violated the ancient agreements of 1447, which had made Bologna free and independent. Thereupon, Vincenzo Beri degli Antoni, a professor of canon and civil law who had lost his university professorship for refusing to swear allegiance to the republic, drafted, at the behest of the city senate, a document aimed at demonstrating the validity of the ancient agreements between Bologna and the Holy See, and the existence, based on the law of nations, of an obligation on the part of the Pope to restore to the city the rights that had been taken away.

Vattel was invoked as the foundation of this theory, and cited in particular was his principle according to which the feared administrative and political reorganisation could not inflict harm on the tradition of autonomy of the city of Bologna, given that the Napoleonic occupation had not been a consequence of the will of the citizens but rather of the Pope's failure to provide protection from the foreigner. Vattel had, in fact, clearly stated that the sovereign had a duty to protect the person and property of his subjects from the enemy, and consequently the lack of such protection constituted in itself a violation of the ancient agreements.[33] The text of Beri degli Antoni initially remained unpublished, only to be surprisingly rediscovered and published twenty years later. The same arguments would in fact be taken up once more in Bologna with the 1831 revolution, when the provisional government declared the temporal power of the Popes to be over, finally publishing for the occasion Berni degli Antoni's text as a historical and political document on the rights of the Bolognese nation.

From Gottingue to the Congress of Vienna: Vattel, Martens and the Survival of the Small States

The use of the *Droit des gens* in Gottingue in the 1770s and 1780s thus allows us to open a window onto the impact of the work on the generation of scholars and law practitioners and international politics during the years of the Napoleonic Empire and then the culture of the Restoration. This was a context profoundly different from that of the initial spread and eighteenth-century discussion of the *Droit des gens*. The years of the French Revolution, but above all those of the Napoleonic Empire and the Restoration, polarised the political debate by permanently transforming in the eyes of critics Vattel's systematisation effort into an ideological commitment no longer compatible with the neutral positions that the author had sought to demonstrate and keep alive.

At the self-same time that Tilliers was producing his abrégé, Gottingue was also the place where one of the scholars who most influenced the law of nations and the theory of diplomacy in the early nineteenth century was undertaking studies: Georg Friedrich von Martens (1756–1821).[34] Having entered the university in 1775, von Martens graduated in 1780 and immediately began work as a lecturer, becoming an associate professor in 1783 and a full professor a year later. Along with his academic career, from the 1790s onwards von Martens began to play a role in diplomacy, which led him, after the fall of Napoleon, to work for the Ministry of State of the principality of Hanover and to join the delegation that took part in the Congress of Vienna.[35] Von Martens, therefore, hailing from a country of limited size like the Electorate of Hanover and being aware of the historical role played by small states in international politics, clearly understood the function and importance of retaining them for the strengthening of the European balance of power and the restoration of the conditions guaranteeing stability that had preceded the Napoleonic hegemony. Moreover, his father, Conrad, had been the Danish consul to Venice from 1739 up to his death in 1786, and his brother Wilhelm Conrad (1748–1828) remained in Venice as a consul, founding the Venetian branch of the family.[36] The study manual that von Martens published in 1785[37] provided early evidence of his indebtedness to the work of Vattel and this was consolidated by the constant strengthening of his reflection on the importance of the voluntary element of the theory of sources of international law.[38]

Through this and other avenues, Vattel and the *Droit des gens* were frequently referred to during the deliberations of the Congress of Vienna, since there was a possibility of reconstructing an international system of European balances which, seen as a deterrent to new despotism, was meant to re-establish the peace guaranteed by the traditional powers of the *ancien régime*. It is well known that Vattel was much used in the rhetoric strategy and the language of the Prussian delegates to the Congress[39] and that the work of Friedrich von Gentz borrowed a great deal from the *Droit des gens*.[40] In the Italian peninsula too the governmental circles of the old Italian states and those of reactionary culture made wide use of the book when speaking or writing in support of the Congress of Vienna and its outcomes.

An example of this is the long *Analisi ragionata del Congresso di Vienna* by the Abbot Paolo Vergani, published in two volumes in 1818.[41] Vergani, who as a young man in the Papal States had sympathised with Enlightenment ideas and engaged in the debate in defence of luxury,[42] now held decidedly reactionary positions, extolling the results of the Congress of Vienna as a product of a collaboration plan between governments and reactionary forces, which triumphed over the subversive continuity represented by a line stretching from the Protestant Reformation to the Enlightenment, Jacobinism, Bonapartism and eventually liberal culture.[43] In the first part of the *Analisi* he took up Vattel's argument apropos of neutrality, especially commercial neutrality, and, following (but not citing) Edmund Burke,[44] he attacked France and its commercial policy. The objective of this was to underline the difference between the conventional law of nations, which for its intrinsic characteristics was variable and subject to changes in the political orientation of nations, and the primitive and universal law of nations, that which is natural, unchangeable and eternal. Thus during the negotiations of the Congress of Vienna neither Napoleonic France nor any of its ancient allies could evoke a standard of conventional international law to nullify a higher principle deriving from the law of nature (the question referred to concerned armed neutrality and the general applicability of the rule according to which a ship's flag covered its cargo, which was to be considered the simple fruit of treaties and conventions).[45]

The reactionary and conservative reading of the *Droit des gens* certainly appears to have held sway in the first two decades of the century precisely because of the system sanctioned by the Congress of Vienna and the Holy Alliance. Moreover, during the 1820s there was a significant increase of new editions of the *Droit des gens*, enough in fact to give rise to talk of a first "Vattel Renaissance."[46] It would nevertheless be wrong to believe that this revival of interest implied a definitive inscription of the work, tout court, into the ideological system of the Restoration. Even in this regard we can identify different stages in the discussion of Vattel's work: the first reading, basically reactionary, was followed in the course of the 1820s (due to the constitutional uprisings of 1820–1821) by a season of intense study by the Italian revolutionary political culture, both that of the Carbonari and constitutional. It was then that the first wave of political exiles and those who joined them after the uprisings, committed to managing the legacy of the Enlightenment culture, perceived in Vattel a proponent of an anti-conservative, liberal international order founded on principles of republican equality.[47]

The phenomenon of the eighteenth-century republications was therefore driven by an agenda, not conservative and largely underground, that brought Vattel's work—along with other classics of eighteenth-century political thought, such as those of Filangieri—into the hands of advocates of liberal constitutionalism to formulate a language of freedom in opposition to the despotism of the Holy Alliance and the political vocabulary imposed by Metternich. Thus Vattel was invoked against the system of balances of power established in Vienna, to propose an alternative, republican system of balance between all states, based on principles of common rights, equality of nations and the containment of excessive influences or new despotisms on the continent. In England, men like the former Jacobin Luigi Angeloni and Gian-Battista Marocchetti used the *Droit des gens* to reopen the debate on the role of the small states (including the kingdom of the Two Sicilies) when confronted by the bigotry of the larger states. Alerino Palma, with his *Difesa dei Piemontesi inquisiti a causa degli avvenimenti del 1821*, utilised the *Droit des gens* to uphold and justify the rights of a people to defend itself against a despotic or absolutist sovereign. Similar positions were taken in Naples in the aftermath of the failed insurrections of 1821 and the intervention of the Spanish army.[48] In Switzerland Franchino Rusca, in the 1830s, turned to Vattel when he defended Swiss freedoms, particularly that of the press, in the face of pressure from Austria and the Kingdom of Sardinia, while the Swiss federal government made open reference to the *Droit des gens* in a formal address to the parliament on the same issue.[49] The Carbonari also availed themselves of the book to theorise the overthrowing of the power relations established by the Congress of Vienna by means of a return to the pre-revolutionary[50] system of balances. Meanwhile Italian liberals exiled in England continued to defend the idea of small states operating within a federal framework by drawing on Vattel's theories to promote a new liberal order between nations willing to come to each other's aid in the name of brotherhood and liberty.[51] A classic example is that of Alerino Palma, a Piedmontese magistrate who in 1822 had fled to London to escape the death sentence before going to Greece in 1826 to take part in the struggle for independence from the Ottoman Empire in a context where Vattel, with Montesquieu and Bentham, was one of the favourite authors.[52]

Even in nineteenth-century Italy, therefore, Vattel's legacy continued to be controversial, not least because it had clearly become a battleground for ideological clashes within rigid interpretative categorisations that would condition the reading of the *Droit des gens* up to the twenty-first century.

NOTES

1. BB, Mss.h.h.X.117, cc. 243. The manuscript occupies a single volume in a hardback binding and contains an aristocratic *ex libris* with the initials "A.T." on its inside front cover.

2. *Genève et la Suisse dans la pensée politique: actes du colloque de Genève (14–15 septembre 2006)*, ed. Christian Poncelet, Rolf Büttiker and Giovanni Busino (Marseille: Presses Universitaires d'Aix-Marseille, 2007); Dino Carpanetto, *Divisi dalla fede. Frontiere religiose, modelli politici, identità storiche nelle relazioni tra Torino e Ginevra (XVII–XVIII secolo)* (Turin: Utet, 2009), 227–279.

3. There is a nod to this in Christoph Good, *Emer de Vattel (1714–1767). Naturrechtliche Ansätze einer Menschenrechtsidee und des humanitäre Völkerrecht im Zeitalter der Aufklärung* (Zurich: Dike Verlag AG, 2011), 52–53.

4. BB, Mss.h.h.X.117, c. 1. The manuscript will hereafter be referred to simply as the Bern manuscript. The numeration of the pages, starting from the text, was made at the same time as the copy and refers to every page. A modern numeration of the blank pages before and after the text has been added in Roman numerals and in pencil.

5. BB, *Bemerkungen einer Reise von Göttingen nach Hamburg und Braunschweig*, Ms. h.h.X.118.

6. On the family and its most important members, see Eduard Blösch, "Johann Anton von Tillier," in *Sammlung bernischer Biographien* vol. 2 (Bern: Francke, 1896), 542–547; Ernst Burkhard, *Johann Anton von Tillier als Politiker* (Bern: Historischer Verein des Kantons Bern, 1963); Richard Feller and Edgar Bonjour, *Geschichtsschreibung der Schweiz*, 2 vols. (Basel: Schwabe, 1979), vol. 2, 612–616.

7. Wolfgang Gresky, "Der Reichsgraf Johann Ludwig von Wallmoden-Gimborn und sein Schlosschen im Georgengarten'," *Hannoverscher Geschichtsblätter* 36 (1982): 269.

8. Georg Christoph Lichtenberg, *Briefwechsel*, ed. Ulrich Joost and Albrecht Schöne, 3 vols. (Munich: Beck, 1983–1985), vol. 3, 160–161.

9. This refers to Emer de Vattel, *Mémoires politiques & militaires pour servir à l'histoire de notre tems*, vol. 2 (Frankfurt and Leipzig: 1760), 386. On the *Mémoires* see Koen Stapelbroek, "The Foundations of Vattel's "System" of Politics and the Context of the Seven Year's War: Moral Philosophy, Luxury and the Constitutional Commercial State," in *The Legacy of Vattel's Droit des gens: Contexts, Concepts, Reception, Translation and Diffusion*, ed. Koen Stapelbroek and Antonio Trampus (Cham: Palgrave Macmillan, 2019), 123.

10. I am referring to the pirate edition of Leiden and that of Lyon published in the same period as the Neuchâtel edition, the partial one published in the *Observateur hollandois* in 1758 (on which see Koen Stapelbroek, "The Foundations of Vattel's Droit des gens," 123–128), and the reprints of 1773 and 1775.

11. See l'*Abrégé de la vie de M. Vattel*, in Emer de Vattel, *Le droit des gens ou principes de la loi naturelle appliqués à la conduite et aux affaires des nations et des souverains. Nouvelle édition augmentée* (Neuchâtel: Société Typographique, 1773), i–vi.

12. In fact the library of the University of Göttingen instead preserves two copies of the pirate edition published in Leiden in 1758, one of which came from another library, as well as the Amsterdam edition of 1775.

13. Letter by Vattel to Malesherbes, 6 December 1757, quoted in Antonella Alimento, "The French Reception of Vattel's Droit des gens: Politic and Publishing Strategies," in Stapelbroek and Trampus, eds., *The Legacy of Vattel's Droit des gens*, 135–164.

14. See the Bern manuscript, c. 221 and Vattel, *Droit des gens*, ed., 1773, book IV, ch. VIII, § 110. The reference is to the *Journal politique* edited in Buillon, which is also known as the *Gazette des gazettes*, 15 January 1771, 50, and 1 February 1771, 54.

15. On the changes under way in the Swiss political culture of the second half of the eighteenth century, also in respect of the legacy of Vattel, Simone Zurbuchen's work should be taken into consideration. See in particular her most recent contributions, "Das Verhältnis Europas zu den Staaten der Alten und der Neuen Welt. Die Idee einer société générale du genre humain in Emer von Vattels Völkerrecht," in *Europa und die Moderne im langen 18. Jahrhundert*, ed. Olaf Asbach (Hanover: Wehrhahn Verlag, 2014), 167–188, and also the long chapter on "Die Schweiz, in Grundriss der Geschichte der Philosophie. Die Philosophie des 18. Jahrhunderts," *Heiliges Römisches Reich Deutscher Nation. Schweiz, Nord- und Osteuropa*, vol. 5, ed. Helmut Holzhey and Vilem Mudroch (Basel: Schwabe, 2014), 1445–1485.

16. Manfred Friedrich, *Geschichte der deutschen Staatsrechtswissenschaft* (Berlin: Duncker und Humblot, 1997), 103.

17. For references to the relationship between *jus gentium* and the science of the state in the teaching in Göttingen see Gabriella Valera, *Scienza dello Stato e metodo storiografico nella scuola storica di Gottinga* (Naples: Edizioni Scientifiche Italiane, 1980), in particular 16, 20 and 108, and also Michael Stolleis, *Public Law in Germany: A Historical Introduction from the 16th to the twenty-first Century* (Oxford: Oxford University Press, 2017), ch. 4. There is also some reference to the relationship between constitutionalism, political history and *jus gentium* in the Göttingen school in Michael C. Carhart, *The Science of Culture in Enlightenment Germany* (Cambridge, MA: Harvard University Press, 2007), 57.

18. *Il diritto delle genti, ovvero Principii della legge naturale, applicati alla condotta e agli affari delle nazioni e de' sovrani. Opera scritta nell'idioma francese dal sig. di Vattel e recata nell'italiano da Lodovico Antonio Loschi*, 3 vols. (Bologna: Fratelli Masi, 1804–1805).

19. On Giuseppe Prina (1775–1859) and his teaching see Ettore Dezza, "Dalle "scienze utili" alle "scientifiche professioni": la formazione universitaria di Giacomo Giovannetti," in Id., *Saggi di storia del diritto penale moderno* (Milan: Giuffrè, 1992), 379–381; Anna Andreoni and Paola Demuru, *La facoltà politico-legale dell'università di Pavia nella Restaurazione (1815–1848)* (Milan: Cisalpino, 1999), 101.

20. Domenico Alberto Azuni, *Dizionario universale ragionato della giurisprudenza mercantile*, vol. 1 (Livorno: Masi, 1822), 330.

21. Domenico Alberto, *Principi del dritto marittimo dell'Europa*, vol. 2 (Trieste: Wage, Fleis & C., 1797), 7.

22. Azuni, *Sistema universale*, vol. 2, 29–30, 38.

23. On Tanucci's view on good government see Roberto Tufano, "Il popolo nel governo di Bernardo Tanucci. L'emergenza della questione sociale nel Regno di Napoli (1734–1774)," in *Un'isola nel contesto mediterraneo.*

Politica, cultura e arte nella Sicilia e nell'Italia meridionale in età medi-evale e moderna, ed. Carmelina Urso, Paola Vitolo and Emanuele Piazza (Bari: Adda, 2018), 103–148; Federico D'Onofrio, "La 'nazione meglio polita': buon governo e costituzione economica della Cina alla scuola di Genovesi," in *Società e storia* 161 (2018): 471–497.

24. Letter of Bernardo Tanucci to Luigi Viviani (27 December 1779) in Enrica Viviani Della Rocca, *Bernardo Tanucci e il suo più importante epistolario* (Florence: Sansoni, 1942), vol. 2, 591–592.

25. Michelangelo Grisolia, *Principj di diritto pubblico ovvero saggio sopra i libri del diritto della guerra e della pace* (Naples: Morelli, 1791), 35–39.

26. Conforti's manuscript was discovered by Fabrizio Lomonaco, *A partire da Giambattista Vico. Filosofia, diritto e letteratura nella Napoli del secondo Settecento* (Rome: Edizioni di Storia e Letteratura, 2010), 195.

27. Specific references in Marc Belissa, *Fraternité universelle et intérêt national (1713–1795). Les cosmopolitiques du droit des gens* (Paris: Kimé, 1998), 384–385.

28. Angela De Benedictis, *Repubblica per contratto. Bologna: una città europea nello Stato della Chiesa* (Bologna: il Mulino, 1995), 380–381; Angela De Benedictis, *Neither disobedients nor rebels: lawful resistence in early modern Italy* (Rome: Viella, 2018).

29. Angela De Benedictis, "Contrattualismo e repubblicanesimo in una città d'antico regime: Bologna nello Stato della Chiesa," in *Materiali per una storia della cultura giuridica* 22 (1992): 269–299.

30. Angela De Benedictis, "Nazione per diritto delle genti: Bologna città libera nello Stato della Chiesa," in *Nazioni d'Italia. Identità politiche e apparte-nenze regionali tra Settecento e Ottocento*, ed. Angela De Benedictis, Irene Fosi and Luca Mannori (Rome: Viella, 2012), 202–204.

31. Francesco Repetti, "Attività editoriale a Livorno tra Settecento e Ottocento: la stamperia di Tommaso Masi," *Nuovi Studi livornesi* 3 (1995): 92–125; Antonio Trampus, "La genesi e la circolazione della Scienza della legislazi-one: saggio bibliografico," *Rivista Storica Italiana* 107 (1) (2005): 336; Maria Grazia Tavoni, "Tipografi, editori, lettura," in *Storia di Bologna*, ed. Aldo Berselli and Angelo Varni, vol. 4/1 (Bologna: Bononia University Press, 2010), 697–698.

32. ASN, Archivio Borbone, b. 108, letter of 10.9.1808. On this journey see Raffaele Feola, *Dall'Illuminismo alla Restaurazione. Donato Tommasi e la legislazione delle Sicilie* (Naples: Jovene, 1982), 99–105. On Vattel's echoes in Spain see Bartolomé Clavero, *Happy Constitution. Cultura y len-gua constitucionales* (Madrid: Editorial Trotta, 1997), 168–173.

33. Vincenzo Berni degli Antoni, *Voto politico-legale per la città di Bologna* (Paris: n.p., 1831). On this text see Angela De Bendictis, "Nazione per diritto delle genti," 210–212.

34. Dietrich Rauschning, *Georg Friedrich von Martens (1756–1821)*. *Lehrer der praktischen Europäischen Völkerrechts und der Diplomatie zu Göttingen*, in *Rechtswissenschaft in Göttingen*. *Göttinger Juristen aus 250 Jahren*, ed. Fritz Loos (Göttingen: Vandenhoeck und Ruprecht, 1987), 123–145.

35. Rauschnigg, *Georg Friedrich von Martens*, 144.

36. Mauro Manfrin, "La famiglia von Martens alla Mira vecchia," *Rive* 8 (2011): 72–79.

37. Georg Friedrich Martens, *Primae lineae iuris gentium Europaerum practici in usum auditorum adumbratae* (Göttingen: Dieterich, 1785).

38. Jean d'Aspremont, *Formalism and the Sources of International Law. A Theory of the Ascertainment of Legal Rules* (Oxford: Oxford University Press, 2011), 64–65.

39. Brian E. Vick, *The Congress of Vienna. Power and Politics after Napoleon* (Cambridge, MA: Harvard University Press, 2014), 307. There is a discussion on this period of change and on the debt that modern international law owes to the law of nations and natural law in Stéphane Beaulac, "Emer de Vattel and the Externalization of Sovereignty," *Journal of the History of International Law* 5 (2003): 237–292.

40. Bruno Arcidiacono, "De la balance politique et des rapports avec les droits des gens: Vattel, la 'guerre pour l'équilibre' et le système européen," in *Vattel's International Law in a XXIst Century Perspective / Le droit International de Vattel vu du XXIe siècle*, ed. Vincent Chetail and Peter Haggenmacher (The Hague: Martinus Nijhoff, 2011), 82.

41. [Paolo Vergani], *Analisi ragionata del Congresso di Vienna*, 2 vols. (Genoa: Stamperia Pagano, 1818).

42. Cecilia Carnino, *Lusso e benessere nell'Italia del Settecento* (Milan: FrancoAngeli, 2014), 237–238.

43. Jörn Leonhard, *Liberalismus: zur historischen Semantik eines europäischen Deutungsmuster* (Munich: R. Oldenbourg Verlag, 2001), 315–316; Nicola Del Corno, *Reazione*, in *Atlante culturale del Risorgimento. Lessico del linguaggio politico dal Settecento all'Unità*, ed. Alberto M. Banti, Antonio Chiavistelli, Luca Mannori and Marco Meriggi (Rome-Bari: Laterza, 2011), 163–167.

44. On Burke's attitude to Vattel, see David Armitage, *Foundations of Modern Political Thought* (Cambridge: Cambridge University Press, 2013), 167–169 and 169. For the consequences of this interpretation of the work of Vattel, see also Koen Stapelbroek, "The Foundations of Vattel's "System" of Politics and the Context of the Seven Years' War: Moral Philosophy, Luxury and the Constitutional Commercial State," in Stapelbroek and Trampus, eds., *The Legacy of Vattel's Droit des gens*, 95–95.

45. [Vergani], *Analisi ragionata*, vol. 1, 89–90.

46. Eugenio Di Rienzo, *Decadenza e caduta del cosmopolitismo: Francia/ Europa, 1792–1848. Note per una ricerca*, in *L'idea di cosmopolitismo: circolazione e metamorfosi*, ed. Lorenzo Bianchi (Naples: Liguori, 2002), 449.
47. Maurizio Isabella, *Risorgimento in Exile: Italian Emigrés and the Liberal International in the Post-Napoleonic Era* (Oxford: Oxford University Press, 2009).
48. Isabella, *Risorgimento in Exile*, 134–135.
49. Fabrizio Mena, *Stamperie ai margini dell'Italia. Editori e librai nella Svizzera italiana 1746–1848* (Bellinzona: Casagrande, 2003), 228.
50. Christopher Alan Bayly and Eugenio Federico Biagini, *Giuseppe Mazzini and the Globalization of Democratic Nationalism 1830–1920* (Oxford: Oxford University Press, 2008), 48–49.
51. Isabella, *Risorgimento in Exile*, 136.
52. Allan Cunningham, *Anglo-Ottoman Encounters in the Age of Revolution*, ed. Edward Ingram, 2 vols. (London: Frank Cass, 1993), vol. 1, 243.

BIBLIOGRAPHY

Most of the quotations of Vattel's *Droit des gens* come from the widely available English-language edition of 2008, which includes an introduction by Béla Kapossy and Richard Whatmore (Indianapolis: Liberty Fund) and maintains the text and English title *The Law of Nations: Or, Principles of the Law of Nature Applied to the Conduct and Affairs of Nations and Sovereigns* of the 1797 London standard edition.

Contemporary translations of primary sources are listed under the names of their authors. Full manuscript sources are referenced only in the notes for reasons of space.

ARCHIVAL ABBREVIATIONS

ASN Archivio di Stato di Napoli, Naples, Italy
BB Bürgerbibliothek Bern, Berne, Switzerland

PRIMARY SOURCES

JOURNALS

Journal politique [*Gazette des gazettes*], 1–15 January 1771 and 1–15 February 1771 (Bouillon: n.p., 1771).
Observateur hollandois ou quarantième lettre de M. van*** à M. H. de la Haye (La Haye: n.p., 1758)

Printed Books

Azuni, Domenico Alberto, *Dizionario universale ragionato della giurisprudenza mercantile*, vol. 1 (Livorno: Masi, 1822)

————, *Principi del dritto marittimo dell'Europa*, 2 vols. (Trieste: Wage, Fleis & C., 1797)

Berni degli Antoni, Vincenzo, *Voto politico-legale per la città di Bologna* (Paris: n.p., 1831)

Grisolia, Michelangelo, *Principj di diritto pubblico ovvero saggio sopra i libri del diritto della guerra e della pace* (Naples: Morelli, 1791)

Lichtenberg, Georg Christoph, *Briefwechsel*, eds. Ulrich Joost and Albrecht Schöne, 3 vols.

Martens, Georg Friedrich, *Primae lineae iuris gentium Europaerum practici in usum auditorum adumbratae* (Gottingue: Dieterich, 1785)

Vattel, Emer de, *Il diritto delle genti, ovvero Principii della legge naturale, applicati alla condotta e agli affari delle nazioni e de' sovrani. Opera scritta nell'idioma francese dal sig. di Vattel e recata nell'italiano da Lodovico Antonio Loschi*, 3 vols. (Bologna: Fratelli Masi, 1804–1805)

————, *Law of Nations: Or, Principles of the Law of Nature, Applied to The Conduct and Affairs of Nations and Sovereigns*, eds. Béla Kapossy and Richard Whatmore (Indianapolis: Liberty Fund 2008)

————, *Le droit des gens ou principes de la loi naturelle appliqués à la conduite et aux affaires des nations et des souverains. Nouvelle édition augmentée* (Neuchâtel: Société Typographique, 1773)

————, *Mémoires politiques & militaires pour servir à l'histoire de notre tems*, vol. 2 (Frankfurt and Leipzig: 1760)

Vergani, Paolo Vergani, *Analisi ragionata del Congresso di Vienna*, 2 vols. (Genoa: Stamperia Pagano, 1818)

Secondary Sources

Alimento, Antonella, "The French Reception of Vattel's Droit des gens: Politic and Publishing Strategies," in *The Legacy of Vattel's Droit des gens*, ed. Koen Stapelbroek and Antonio Trampus (Basingstoke: Palgrave Macmillan, 2019), 135–164

Andreoni, Anna and Demuru, Paola, *La facoltà politico-legale dell'università di Pavia nella Restaurazione (1815–1848)* (Milan: Cisalpino, 1999)

Arcidiacono, Bruno, "De la balance politique et des rapports avec les droits des gens: Vattel, la 'guerre pour l'équilibre' et le système européen," in *Vattel's International Law in a XXIst Century Perspective / Le droit International de Vattel vu du XXIe siècle*, ed. Vincent Chetail and Peter Haggenmacher (The Hague: Martinus Nijhoff, 2011), 77–100

Armitage, David, *Foundations of Modern Political Thought* (Cambridge: Cambridge University Press, 2013)

Aspremont, Jean d', *Formalism and the Sources of International Law. A Theory of the Ascertainment of Legal Rules* (Oxford: Oxford University Press, 2011)

Bayly, Christopher Alan and Biagini, Eugenio Federico, *Giuseppe Mazzini and the Globalization of Democratic Nationalism 1830–1920* (Oxford: Oxford University Press, 2008)

Beaulac, Stéphane, "Emer de Vattel and the Externalization of Sovereignty," *Journal of the History of International Law* 5 (2003), 237–292

Belissa, Marc, *Fraternité universelle et intérêt national (1713–1795). Les cosmopolitiques du droit des gens* (Paris: Kimé, 1998)

Blösch, Eduard, *Johann Anton von Tillier*, in *Sammlung bernischer Biographien* vol. 2 (Bern: Francke, 1896), 542–547

Burkhard, Ernst, *Johann Anton von Tillier als Politiker* (Bern: Historischer Verein des Kantons Bern, 1963)

Carhart, Michael C., *The Science of Culture in Enlightenment Germany* (Cambridge, MA: Harvard University Press, 2007)

Carnino, Cecilia, *Lusso e benessere nell'Italia del Settecento* (Milan: FrancoAngeli, 2014)

Carpanetto, Dino, *Divisi dalla fede. Frontiere religiose, modelli politici, identità storiche nelle relazioni tra Torino e Ginevra (XVII–XVIII secolo)* (Turin: Utet, 2009)

Clavero, Bartolomé, *Happy Constitution. Cultura y lengua constitucionales* (Madrid: Editorial Trotta, 1997)

Cunningham, Allan, *Anglo-Ottoman Encounters in the Age of Revolution*, ed. Edward Ingram, 2 vols. (London: Frank Cass, 1993)

D'Onofrio, Federico, "La 'nazione meglio polita': buon governo e costituzione economica della Cina alla scuola di Genovesi," in *Società e storia* 161 (2018), 471–497

De Benedictis, Angela, "Contrattualismo e repubblicanesimo in una città d'antico regime: Bologna nello Stato della Chiesa," in *Materiali per una storia della cultura giuridica* 22 (1992), 269–299

——, "Nazione per diritto delle genti: Bologna città libera nello Stato della Chiesa," in *Nazioni d'Italia. Identità politiche e appartenenze regionali tra Settecento e Ottocento*, eds. Angela De Benedictis, Irene Fosi and Luca Mannori (Rome: Viella, 2012), 195–216

——, *Neither disobedients nor rebels: lawful resistance in early modern Italy* (Rome: Viella, 2018)

——, *Repubblica per contratto. Bologna: una città europea nello Stato della Chiesa* (Bologna: il Mulino, 1995)

Del Corno, Nicola, *Reazione*, in *Atlante culturale del Risorgimento. Lessico del linguaggio politico dal Settecento all'Unità*, eds. Alberto M. Banti, Antonio

Chiavistelli, Luca Mannori and Marco Meriggi (Rome-Bari: Laterza, 2011), 163–167

Dezza, Ettore, "Dalle "scienze utili" alle "scientifiche professioni": la formazione universitaria di Giacomo Giovannetti," in *Saggi di storia del diritto penale moderno* (Milan: Giuffrè, 1992), 367–387

Di Rienzo, Eugenio, *Decadenza e caduta del cosmopolitismo: Francia/Europa, 1792–1848. Note per una ricerca*, in *L'idea di cosmopolitismo: circolazione e metamorfosi*, ed. Lorenzo Bianchi (Naples: Liguori, 2002), 419–458

Feller, Richard and Bonjour, Edgar, *Geschichtsschreibung der Schweiz*, 2 vols. (Basel: Schwabe, 1979)

Feola, Raffaele, *Dall'Illuminismo alla Restaurazione. Donato Tommasi e la legislazione delle Sicilie* (Naples: Jovene, 1982)

Friedrich, Manfred, *Geschichte der deutschen Staatsrechtswissenschaft* (Berlin: Duncker und Humblot, 1997)

Good, Christoph, *Emer de Vattel (1714–1767). Naturrechtliche Ansätze einer Menschenrechtsidee und des humanitäre Völkerrecht im Zeitalter der Aufklärung* (Zurich: Dike Verlag AG, 2011).

Gresky, Wolfgang, "Der Reichsgraf Johann Ludwig von Wallmoden-Gimborn und sein Schlösschen im Georgengarten'," *Hannoverscher Geschichtsblätter* 36 (1982), 252–279

Isabella, Maurizio, *Risorgimento in Exile: Italian Emigrés and the Liberal International in the Post-Napoleonic Era* (Oxford: Oxford University Press, 2009)

Leonhard, Jörn, *Liberalismus: zur historischen Semantik eines europäischen Deutungsmuster* (Munich: R. Oldenbourg Verlag, 2001)

Lomonaco, Fabrizio, *A partire da Giambattista Vico. Filosofia, diritto e letteratura nella Napoli del secondo Settecento* (Rome: Edizioni di Storia e Letteratura, 2010)

Manfrin, Mauro, "La famiglia von Martens alla Mira vecchia," *Rive* 8 (2011), 72–79

Mena, Fabrizio, *Stamperie ai margini dell'Italia. Editori e librai nella Svizzera italiana 1746–1848* (Bellinzona: Casagrande, 2003)

Poncelet, Christian, Büttiker Rolf, and Busino, Giovanni, eds., *Genève et la Suisse dans la pensée politique: actes du colloque de Genève (14–15 septembre 2006* (Marseille: Presses Universitaires d'Aix-Marseille, 2007)

Rauschning, Dietrich, *Georg Friedrich von Martens (1756–1821). Lehrer der praktischen Europäischen Völkerrechts und der Diplomatie zu Göttingen*, in *Rechtswissenschaft in Göttingen. Göttinger Juristen aus 250 Jahren*, ed. Fritz Loos (Göttingen: Vandenhoeck und Ruprecht, 1987), 123–145

Repetti, Francesco, "Attività editoriale a Livorno tra Settecento e Ottocento: la stamperia di Tommaso Masi," *Nuovi Studi livornesi* 3 (1995), 92–125

Stapelbroek, Koen, "The Foundations of Vattel's "System" of Politics and the Context of the Seven Year's War: Moral Philosophy, Luxury and the Constitutional Commercial State," in *The Legacy of Vattel's Droit des gens:*

Contexts, Concepts, Reception, Translation and Diffusion, eds. Koen Stapelbroek and Antonio Trampus (Cham: Palgrave Macmillan, 2019, 95-123

Stolleis, Michael, *Public Law in Germany: A Historical Introduction from the 16th to the 21st Century* (Oxford: Oxford University Press, 2017)

Tavoni, Maria Grazia, "Tipografi, editori, lettura," in *Storia di Bologna*, eds. Aldo Berselli and Angelo Varni, vol. 4/1 (Bologna: Bononia University Press, 2010), 687–868

Trampus, Antonio, "La genesi e la circolazione della 'Scienza della legislazione'. Saggio bibliografico," *Rivista Storica Italiana* CXVII (2005), 309–359

Tufano, Roberto, "Il popolo nel governo di Bernardo Tanucci. L'emergenza della questione sociale nel Regno di Napoli (1734–1774)," in *Un'isola nel contesto mediterraneo. Politica, cultura e arte nella Sicilia e nell'Italia meridionale in età medievale e moderna*, eds. Carmelina Urso, Paola Vitolo and Emanuele Piazza (Bari: Adda, 2018), 103–148

Valera, Valera, *Scienza dello Stato e metodo storiografico nella scuola storica di Gottinga* (Naples: Edizioni Scientifiche Italiane, 1980)

Vick, Brian E., *The Congress of Vienna. Power and Politics after Napoleon* (Cambridge, MA: Harvard University Press, 2014)

Viviani Della Rocca, Enrica, *Bernardo Tanucci e il suo più importante epistolario* vol. 2 (Florence: Sansoni, 1942)

Zurbuchen, Simone, "Das Verhältnis Europas zu den Staaten der Alten und der Neuen Welt. Die Idee einer société générale du genre humain in Emer von Vattels Völkerrecht," in *Europa und die Moderne im langen 18. Jahrhundert*, ed. Olaf Asbach (Hanover: Wehrhahn Verlag, 2014a), 167–188

———, "Die Schweiz, in Grundriss der Geschichte der Philosophie. Die Philosophie des 18. Jahrhunderts," *Heiliges Römisches Reich Deutscher Nation. Schweiz, Nord- und Osteuropa*, vol. 5, ed. Helmut Holzhey and Vilem Mudroch (Basel: Schwabe, 2014b), 1445–1485

State and Nation: The Political Neutralisation of the *Droit des gens* in Nineteenth-Century Europe

Following the Restoration, the political culture of nineteenth-century Europe found itself making use of the *Droit des gens* while viewing it from two different but complementary standpoints. On the one hand, there were those who read the text as it had been written by Vattel, with his typically eighteenth-century language and concepts attached to interpretative systems belonging to the culture of natural law and the Enlightenment. On the other hand, the intellectuals of the Restoration were also well aware of the interpretations that had been made of Vattel's work in the age of the Atlantic revolutions, in particular the readings that had transformed it into a dangerous and 'revolutionary' text, when it had been invoked to call into question the sovereignty and principles of authority that supported the society of the *ancien régime*.

A large part of the new editions and studies of the *Droit des gens* in the first half of the nineteenth century were therefore devoted to commenting on and criticising Vattel's text.[1] As a result, in the climate of the Congress of Vienna, of the Restoration and then of the liberal revolutions, it became more and more necessary and urgent to explain, clarify and define the meaning and ideas of the Swiss author. Moreover, increasingly commentators realised that a huge campaign was needed to neutralise the political use of this text, although, in the opinion of contemporaries, there were only two

Archival Abbreviations
BPRN   Biblioteca del Palazzo Reale di Napoli, Naples, Italy

ways in which this could be done effectively: by historicising the *Droit des gens*—in other words by delimiting its value to the historical period in which it was written—and by relegating it from the status of a politico-philosophical text to that of a practical-juridical manual.

FROM MARTENS TO RANKE: LARGE STATES VERSUS SMALL STATES IN THE NEW POWER POLITICS

The large number of new editions and translations of the *Droit des gens* published during the nineteenth century therefore comprised only a partial indication of the real success of Vattel's work. In other words, the success of Vattel's work was not brought about by a desire to disseminate his text, but rather, paradoxically, by an ever-more pressing need to circumscribe and curb that success. While keeping this in mind, we must first reflect on the cultures and geographical and political areas from which most of the new editions and commentaries came. During the course of the eighteenth century the bulk of the readings, commentaries and translations of the *Droit des gens* came—as we have already seen—from the small European states, which had found that Vattel's work provided them with the toolkit they needed for asserting their dignity and sovereignty within the system of balances that ensued from the Treaty of Utrecht. Now, however, the new editions and commentaries came mostly from the cultural circles of the great powers, that is, from the protagonists and arbiters of the new system of European balances established by the strategy of Metternich. France thus came to be at the forefront of a criticism of Vattel which, without disclaiming the relevance of the work, emphasised the need to update it, interpret it and explain it in the light of a context completely different from the one in which it had been written. But even in the Germanic Empire, which had begun to take the form of the Germanic Confederation, the approach was similar. In all these cases the operation was carried out not by intellectuals and men of letters interested in debating the relevance of the *Droit des gens* in support of internal social reforms and the arrangement of their states, but rather by diplomats and statesmen concerned with observing the foreign policy guidelines of their countries on the international stage.

The Martens dynasty played a key role in the transformation of the *Droit des gens* from a politico-philosophical work to a practical manual, and also in the neutralisation of Vattel's theories. As many biographers have noted, Georg Friedrich Martens had grown intellectually and scientifically in the second half of the eighteenth century, learning from Vattel, both as the author of the *Droit des gens*, considered a classic of the law of nations, and as a councillor to the court of Saxony.[2] His indebtedness to Vattel can be discerned in many pages of his *Primae lineae iuris gentium Europaeum practici in usum auditorum adumbratae* (1785),[3] which later became the *Précis du droit des gens moderne de l'Europe* (1788),[4] as well as in his *Einleitung in das positive Europäische Völkerrecht* (1796).[5]

Although it has been observed that in the history of international law the position of Martens does not appear to have been against natural law but instead represented an evolution of it,[6] it should be noted that his cultural orientation—of admiration but also criticism of the natural law tradition—is evident from the moment that he entitled his work *das positive Europäische Völkerrecht* and then in French *droit des gens moderne de l'Europe*. In his opinion, the law of nations was relevant as a positive and modern system, not as an expression of an older tradition. Martens was more interested in the law of nations in force—the one actually implemented by laws and international treaties—than the philosophical system of natural law. The adjective 'modern'—which has only a vague and circumstantial connection with the German 'positive'—was clearly used as the antonym of 'ancient,' thus marking a clear break with the tradition that preceded Vattel.[7]

The Martens family is the missing link that allows us to understand how the historicisation process of the *Droit des gens* took its place in the new international context of the power politics inaugurated by Metternich. As well as their villa in the countryside of Mira, the family had a house in Venice, in Rio di san Cancian, from the time of Conrad's and then Wilhelm Conrad's service as Danish consuls to Venice in the final years of the Republic. The grandson, Karl von Martens, was the son of Wilhelm Conrad and thus the nephew of Georg Friedrich and was a significant figure because he would eventually take up his uncle's work, and would play host to Leopold von Ranke in Venice. The approach taken by Karls (from here on Charles, as he signed his name from the 1820s onwards) to the law of nations would, as we shall see, partly reflect his sensitivity to history and the commitment he shared with Ranke to the recovery of diplomatic sources in the Venetian archives.[8]

Charles Martens made two key contributions to the history of the reception of the *Droit des gens*. It was he who was given the legacy of Georg Friedrich and administered it through the re-editions and summaries of his writings. In addition, by virtue of his diplomatic work—he in fact became foreign commissioner to the King of Prussia—he promoted the use of Vattel's work as a simple deposit of historical-diplomatic information rather than as an international politico-philosophical project. His friendship with Leopold von Ranke was conducive to this operation. The link between the two has never been closely studied but it is nonetheless important for comprehending not only the transformation of the *Droit des gens* into a source of historical-diplomatic information, but also for grasping how a series of cultural connections that would be used during the nineteenth century was activated. Through this path, Vattel became a source to use when studying, for example, the historical role of the Church of Rome in the construction of the ancient architecture of alliances and, in the drafting of treaties, to fathom the system of Westphalia, and to evaluate the historical function performed by the Catholic Church and the Pope.[9]

To understand how Ranke's work intersected with readings of Vattel, it is necessary to return to the years of the German historian's early education. As we know, his interest in the past was not the mere curiosity of an erudite man, and the years of his education in the philological school of Leipzig should be read in relation to his later studies in Berlin. History, for Ranke, was an 'ideal' story, that is, one criss-crossed with cultural and spiritual tensions. The study of antiquity illuminated the modern era, just as the study of great empires of the past threw light on the history of recent great 'empires,' such as the Germanic one or the papacy.

In the years in which, after the Restoration, Ranke applied the philological method to historical criticism, he witnessed the emergence of the Europe of nations and of new great powers, which appeared to him the manifestation of the absolute in the finiteness of the individual. He believed the study of the history of nations, and of German ones specifically, to be essential to understanding the history of humanity, of the generations and of peoples. It must be recognised that "the particular holds within itself the universal" and thus every nation must be traced back to a time that is both absolute and finite, irrational and earthly, arbitrary and necessary.[10]

It was against this background that Ranke laid the foundations for his essay on the great powers (*Die grossen Mächte*, 1833), which contained a radical reformulation of the concept of the 'great power,' and which would have a significant influence on the culture of the nineteenth and twentieth centuries. As with most statesmen of the time, Ranke also held that the only great powers were those that—thanks to their economic and military strength—were able to create a sphere of influence with which the other states had to maintain a *droit de regard*, the right to exercise control. Ranke's distance from the work of Vattel, and from eighteenth-century culture in general, became unbridgeable from this point: to Vattel's mind power (*puissance*) was a quality commensurate with the exercise of political power, with its effectiveness and interest, and was much closer to the terminology of the reason of state than the actual measure of influence exerted on the interstate level.[11] For Vattel all states, large and small but operating inside a logic of formal equality, could therefore play a role within the logic of balance and exercise their power. For Ranke, however, this vision was no longer current and the great powers in their historical mission were incompatible with the pluralism of Vattel's international system.[12]

This reasoning could also be used to analyse the historical events of the ancient states and in particular of the Republic of Venice, which in the second half of the eighteenth century—by using the *Droit des gens*, as we saw in the seventh chapter—had wanted to establish an opposing principle, namely that even a small republican-based state, now without effective military or economic power, could have the dignity to place itself in the international forum on the same level as a great power, provided that its internal organisation and its policy of good government satisfied the criteria of a constitutional state that allowed formal equality between nations. Ranke had been fascinated by Venetian history since his youth and, with the aim of gaining a grasp of the history of the greatness and decadence of the Republic, he was among the first to take advantage of the opening of the *Serenissima*'s archives, which was made possible by the Austrian government after the definitive cession of Veneto to Austria. He first went to Vienna, where he obtained, through Friedrich von Gentz, an audience with Prince Metternich, and then, with the necessary permissions to access the Venetian archives,[13] he arrived in Venice where he was welcomed by the Martens family. It was his studies of these archives and of the history of the republic, later merged into a long series of publications starting in 1831, that convinced him that Venice could have been a power in the sense intended by Vattel, but not a 'great power' in the international political sense, especially after it had ceased to be faithful to its own institutions and tried instead to imitate those of others.

FROM GEORG FRIEDRICH TO CHARLES DE MARTENS

As we have seen, there was a direct link between the de Martens family and Venice and between Ranke and the Martens family. And the course of events affecting Vattel's work during the nineteenth century and its new function as a practical tool for the study of diplomatic history was linked to these actors.

From the time of the *Précis du droit moderne de l'Europe fondé sur les traités et l'usage* (translated into English in 1795) Georg Friedrich Martens had engaged in direct dialogue with the leading authors of public and natural law of the eighteenth century, from Burlamaqui to Mably and Vattel himself. Indeed, the *Droit des gens* was one of the main sources utilised for the *Précis*, for documentary and interpretative purposes.[14] After the *Précis*, which enjoyed a long and uninterrupted period of success lasting several decades, Martens had written another two volumes which contained a collection of well-known cases of modern international law (*Erzählungen merkwürdiger Fälle des neutre Europäischen Völkerrrecht*, Göttingen 1800 and 1802) selected for explanatory reasons to cover a period of time of about fifty years running from the War of the Austrian Succession to 1799.

As the formulation of the project and the introduction to the collection of cases made clear, the two works were to be read together, being complementary: to be specific, the *Précis*, which still followed Vattel's approach, provided the key to the second. In the new context of the Restoration this connection was, however, no longer useful, and the death of Georg Friedrich in 1821, only a few years after having been appointed representative of the King of Hanover to the diet of the Germanic Confederation of Frankfurt, gave impetus to the notion of separating the case studies from their eighteenth-century and natural law interpretative matrix.

Charles de Martens, Georg Friedrich's nephew, was resident minister of the Grand Duchy of Weimar in Dresden. Shortly after his uncle's death, he began to rework his entire corpus, first publishing a *Manuel diplomatique* (1823) and shortly after two volumes of *Causes célèbres du droit des gens* (1827).[15]

With the *Manuel diplomatique* de Martens shifted the public's attention from the eighteenth-century primacy of natural law and the law of nations, intended as foundational rights of the state and as a set of natural principles to be used in harmonising the conduct of nations and directing it towards international law conceived as a living and positive law, which is

to say originated, modified and composed of cases and practical explanations. At the same time, he launched another project, namely that of
reworking and transforming the two volumes of his uncle's *Erzählungen*.
The resulting *Causes célèbres du droit des gens*, published in co-editions by
the printers Brockhaus of Leipzig and Ponthieu of Paris, were in fact an
update of his uncle's work in appearance only. Although the introduction
stated that the author had taken up his uncle's ideas,[16] the work was actually profoundly different in approach. The cases presented twenty years
earlier were reduced in number and abridged, while the timespan under
review was extended to more than a century, beginning in 1703. The
selected cases were therefore no longer representative of a 'modern' law of
nations, in the current sense, but of a historical conception of law and
diplomacy. Moreover, the two volumes were dedicated, significantly, to
the leader of a great power, Nicolas I, Emperor of Russia and King of
Poland, and the great majority of the chosen cases were examples of competition and confrontation between small and large states, and old and
new powers: Savoy against France, Sweden against Great Britain, Spain
against Great Britain, Portugal against Spain, the United Provinces against
France, and so on. The references to Vattel were reduced to a few irrele-
vant citations, which completely weakened the importance and value
which had been attached to the *Droit des gens* in the previous century.

FROM CONSTITUTIONAL MONARCHY TO THE NATION

The two Martens constituted a point of reference for another author who
perhaps was the first to explicitly associate criticism of Vattel's incomplete-
ness and generality with the problem of the nation. He was Silvestre
Pinheiro Ferreira. Like de Martens, Pinheiro was not a jurist or a theorist
but primarily a diplomat, although as a young man he had taught philosophy for a brief time in Coimbra. In the service of the King of Portugal, he
had been secretary of the legation to Paris in 1797 and then followed King
John VI to Brazil, becoming a trade adviser and secretary of state. A supporter of constitutional and representative monarchy, after the Porto
Revolution of 1821 he became foreign and war minister and resigned in
protest against the absolutist regime of Miguel de Braganza, King of
Portugal from 1828 to 1834, thereafter living in exile in Paris, never to
return home.

In this intellectual and international context Pinheiro published in Paris
in 1838 a commentary added to a new edition of the *Droit des gens* by the

publisher Aillaud,[17] who already in 1830 had reprinted the work, stating that it had been published in Paris and Rio de Janeiro. Jean-Pierre Aillaud (1787–1852) was a bookseller and printer specialising in the publication of works by political exiles, in particular the Portuguese. He was born in Coimbra, the son of a French bookseller who had settled there. After opening a bookshop with a publishing house in Paris in 1820, he soon began to specialise in the publication of books on diplomatic and political themes, calling his bookshop the "Librairie diplomatique, français et etrangère." He also became a Portuguese vice-consul in Caen. Aillaud's close links with Portugal and with the circles supportive of the constitutional monarchy in the country, and therefore also with Pinheiro, thus represent a first point of reference for contextualising the re-edition of Vattel's work and Pinheiro's commentary. However, other elements to be taken into consideration are the fact that Aillaud was the publisher of choice also for Neapolitan exiles in Paris and for those like Francesco Saverio Salfi who sought to keep alive the memory of Gaetano Filangieri (Aillaud republished the *Scienza della legislazione*). In addition, the role of Benjamin Constant should also be remembered, since he had already published in 1829 his *Oeuvres diverses sur la politique constitutionelle* with Aillaud.

These references make it easier to understand the motivations and the meaning of Pinheiro's comments on the *Droit des gens*, which would have an impact on the fortunes of Vattel in the nineteenth century. On the one hand, the Portuguese diplomat repeated the observations made by the Martens, especially the claim that the *Droit des gens* should be read in the context of the eighteenth century, that it was no longer topical, and that it needed to be historicised because both the constituting principles of governments and the trade and international relations between nations had changed over the years. On the other hand, Pinheiro introduced an argument typical of nineteenth-century critique of Vattel, namely that the concepts of the *Droit des gens* were too generic, vague, indistinct and erroneous, and that they required greater definition and clarification.[18]

Pinheiro's working method was thus rather curious: he stated that Vattel's work was no longer of use in the Europe of the Restoration, that it had to be irretrievably consigned to history, and that every contemporary reading of it would increase its threat. And yet, instead of ignoring the *Droit des gens*, he acknowledged that it was a classic of modern thought, whose centrality was greater than the works of Wolff and Grotius, and therefore it could not be forgotten but ought to be read, analysed and discussed. This was the same way of proceeding and a similar kind of criticism adopted ten years earlier by Benjamin Constant in regard to Gaetano Filangieri's *Scienza della legislazione*. In that case too the criticism of the eighteenth-century work recognised the central position that it had acquired in the European culture of the late *ancien régime*.[19] It therefore seems very likely that Pinheiro had read Constant and his *Commentaire* of Filangieri.[20] But in any case it is clear that he acted as an interpreter of a cultural context, that of the supporters of constitutional monarchies and liberal constitutions, which he believed had to settle accounts with the classical authors of eighteenth-century culture held largely responsible for the excesses caused by the French Revolution.

According to Pinheiro, therefore, in order to make use of the *Droit des gens* it was no longer enough to create a commentary, but rather a thorough revision of Vattel's text and theories was necessary. The points given most attention had to be those relating to the relationship between state and nation and the wrong definition of what a state was because, according to Pinheiro, the nation was such only when it had the strength to respect and be respected by other nations and states. Another point for discussion was the vagueness with which Vattel, in 1758, had defined the constitution as a fundamental regulation of the state.

For Pinheiro—who at this juncture used an entirely anti-historical line of reasoning by choosing to ignore the context in which Vattel had written his work and the still general meaning attributed in the mid-eighteenth century to the word 'constitution'—the existence of at least three types of law needed to be spelt out more carefully: the actually fundamental one identified with the declaration of natural rights; that of the constitution, useful for establishing the duties of those who exercise political power; and finally the set of organic laws aimed at regulating the exercise of these duties. These different levels and this different ranking also determined the possibility of modifying the fundamental laws and the extent to which this could be done.

Finally, it is interesting to note that through this analysis the logic with which the *Droit des gens* was read during the eighteenth century was definitively abandoned, that is, that of considering it an instrument for the legitimisation of small states on the international scene. The attention of the critics focused on the first book of the work, which was dedicated to the constitutional structure and internal organisation of the state, albeit no longer in the eighteenth-century cosmopolitan sense but in line with the logic of the nation state. This change of direction is even more significant if we consider that Pinheiro's annotations would also be incorporated in the last and definitive nineteenth-century French edition of the *Droit des gens*, the one published in 1863 by Paul Pradier-Fodéré with the insertion of the *Discourse on the study of the laws of nature* by James Mackintosh (1765–1832).[21]

The *Droit des gens* interpreted by Pinheiro had an immediate and widespread diffusion, first of all in the small states of the German Confederation, where the formation process of the nation state was already under way. As early as 1839 the *Literarische Zeitung* and then Leopold August Warnkönig in the *Rechtsphilosophie als Naturlehre des Rechts*[22] had shown an interest in it. In the Italian peninsula the *Annali universali di statistica*, a journal with which jurists and philosophers like Melchiorre Gioia and Giandomenico Romagnosi collaborated, published a long and enthusiastic review in which Pinheiro's commentary was called the expression of the "law of honest people" compared to the ancient law of Vattel.[23] The 1873 *Enciclopedia giuridica* presented Pinheiro's work as one "with the authority of a diplomatic manual."[24] The work received similar attention in the *Lezioni di diritto costituzionale* by Ludovico Casanova[25] and again in the 1880s in the *Trattato di diritto internazionale pubblico* by Pasquale Fiore, a professor at the University of Naples, who thought that Pinheiro and Pradier-Fodéré's notes had "rejuvenated" Vattel, and therefore recommended the *Droit des gens* to "modern science enthusiasts."[26] In this way a new European renaissance of Vattel's work began, one based on interpretations very different from those that had been used in the previous century, but which were curiously symmetrical to the attention given in the United States to the *Droit des gens* (in the Pinheiro version) by a specialist in relations between municipal law and constitutional law, John Norton Pomeroy, a professor at New York University.[27]

VATTEL, THE PROBLEM OF THE NATION AND THE POLITICAL PROCESSES OF 1848

The need to neutralise the political implications of the *Droit des gens* became even more evident during the 1840s and 1850s, as shown by the new way in which the concept of nation was used, above all in continental Europe and in the regions where the Restoration had consolidated systems of government increasingly perceived as foreign and oppressive. A somewhat paradigmatic case is that of the Guerrazzi-Romanelli trial in which good use was made of Vattel's work.

In 1848, when a widespread revolutionary wave swept over Europe, even the Grand Duchy of Tuscany found itself under pressure from democratising and liberalising forces. A popular revolt in Livorno in the spring of that year forced the Grand Duke Leopold II of Habsburg-Lorraine, despite having conceded a constitution, to flee and take refuge in Gaeta—a town in papal territory—where he was given protection by the Austrian forces. In Florence, in February 1849 a provisional government was established under a triumvirate composed of Francesco Domenico Guerrazzi, Giuseppe Montanelli and Giuseppe Mazzoni, and it remained in charge until the Grand Duke's government was restored the following July.

After the return of the Austrians the members of the provisional government and their collaborators were arrested and tried. Among them was Leonardo Romanelli, a childhood friend and fellow student of Guerrazzi, who had been called to the provisional government as minister of justice. Expelled from the Grand Duchy after July 1849, he found refuge in Umbria, another territory of the Pope, but was captured and put into preventive detention on charges of lese-majesty and high treason. He was one of the co-defendants in the trial against Guerrazzi and the triumvirate but, after four years of detention, was acquitted and freed in 1853.[28]

The lawyer Adriano Mari was Romanelli's defence counsel. A liberal, former Tuscan deputy and then deputy to the constituent assembly of 1849, Mari adhered to a moderate position throughout the time of the provisional government and that of the Lorraine restoration, distinguishing himself in the defence of the patriots even when not sharing their more radical views. The defence and exoneration of Romanelli thus became a political issue and Mari's legal victory was greeted with great fanfare. As the court documents demonstrate, Mari's line of defence was essentially that of undermining, on a legal level, the link between state and nation to demonstrate that the interests of the Tuscan Lorraine state did not necessarily coincide with those of the Tuscan nation. Paradoxically, he thus turned to the accused's advantage Vattel's principle that took issue with the Italian Risorgimento's propensity to equate state with nation. Mari's aim was to show beyond doubt that Romanelli, having been banished from Tuscan territory at the time of his arrest, could no longer be considered a subject of the Grand Duchy, and that because of this expulsion, had in fact reacquired that form of natural freedom which existed before the formation of any political bond with the territory of the state. Invoking Vattel, Mari explained that there was no identity of nation with state of the *ancien régime*, and the link between the nation and the state could be dissolved precisely because it was voluntary. The relationship between citizen and state did not originate from the laws of nature, in which instead only the foundations of the nation lay. The state was the result of a political agreement, of "a particular communion of men reunited in a civil consortium,"[29] but above it the law of nations reigned universally. In consequence, the powers of the state—and here Mari's liberal viewpoint emerged—were subject to limits and restraints in the laws of nature, and the state could not behave towards individuals, who were naturally free, like a master towards a slave.

It was therefore necessary, Mari argued, to investigate closely and thoroughly the connection uniting a specific individual to the state to which he belonged. This was in fact a bond that could dissolve, a link that could break, and when this separation occurred the sovereignty of the state ceased, and so too—he said, quoting Vattel—did the man's duties to the state. "Every man is born free," Vattel had written, "and the son of a citizen, when come to the years of discretion, may examine whether it be convenient for him to join the society for which he was destined by his birth."[30]

In this particular case, Romanelli had been banished by the Lorrainese-Austrian government, and was given a passport as a result of his expulsion. These acts made clear that the Tuscan state had wanted to release him from his obligations to the Grand Duke, thus losing and renouncing—in keeping with Vattel's approach—any authority over him. Driven from Tuscany, Romanelli had regained his natural freedom and the Lorrainese government could no longer exercise any power over him, either derived from *imperium*—construed as power and right to regulate the government of the country—or from *dominium*, that is, territorial control.[31] As Vattel had written, it was *imperium* together with *dominium* that qualified sovereignty and formed the jurisdiction over the country in which one lived.[32] It therefore followed that Marinelli had been legitimately removed from any jurisdiction of the Tuscan state.[33]

Vattel, Mari continued in his defence of Marinelli, had also demonstrated that a state (Tuscany, in this instance) was obliged to refrain from performing any act that could offend the sovereignty of another state, and thus the power of the Tuscan state inevitably ended at its borders: the arm of the law could not stretch any further, nor could an exiled individual be followed or captured in the territory of another state. As Vattel had stated, "we cannot then, without doing an injury to a state, enter its territories with force and arms in pursuit of a criminal, and take him from thence."[34]

The subject of exile, to which Vattel had paid specific attention,[35] and which was becoming relevant again in Restoration Europe, thus served as an argument in favour of man's natural freedom: "a man, by being exiled or banished, does not forfeit the human character, nor consequently his right to dwell somewhere on earth," wrote Vattel, as quoted by Mari.[36] An exile, therefore, could not be persecuted in the country where he had taken refuge for crimes committed elsewhere, nor could the host state punish him for crimes or violations not committed in its own territory.[37] Nor, finally, could Tuscany seek the extradition of the guilty party, since the same principles of Vattel and of the law of nature prevented the arrest and handing over of a man who, even if guilty, was only guilty of crimes committed on foreign soil.

Compared with the trials of other members of the provisional government, the one against Romanelli may have attracted more attention precisely because Mari's arguments were accepted and led to an acquittal in July 1853: Romanelli was one of very few of the thirty-six defendants to be cleared.

The interpretations and application of Vattel's work during the first half of the nineteenth century thus also reveal a progressive conceptual and cultural separation of natural law from the positive law of states, which could be used both positively—as in the case of the trial against Guerazzi and Romanelli—and negatively. The *Droit des gens* was already seen by that time as the deposit of a legal and philosophical patrimony connected to a natural law that was turning out to be increasingly incompatible with the positive law of states, which could be invoked and used to ensure that those principles prevailed over the domestic law of the nation states, but at other times was felt to be irredeemably foreign to a continuously more accentuated trend towards the nationalisation of law and politics.

STATE, NATION AND HOMELAND

This phenomenon can be confirmed straight away by means of the transformations in the political language and by means of the attribution of new meanings to key words and concepts widely used by Vattel in the *Droit des gens* a century earlier. By the fourth and fifth decades of the nineteenth century, words like 'state,' 'nation' and 'homeland,' which recurred frequently in Vattel's text, had taken on completely different meanings in the European public debate. On the one hand, then, the *Droit des gens* again became defiant in the face of readings that sought to update its contents; on the other hand, paradoxically, many authors lamented the generic quality of the expressions employed by Vattel and his inability to notice even in his own time the shades of meaning that those words had started to acquire.

An example of how, in the early decades of the nineteenth century, a kind of cultural short circuit was developing around Vattel's work can be found in the lessons of natural and public law held at the University of Pavia by Pietro Baroli and published in 1837.[38] Baroli (1797–1878) was a native of Cremona and, in 1828, had already made a name for himself—when he was a high school teacher in Como—as a critical reader of Kant's philosophy. He went on to become Professor of Philosophy and Public Law at the University of Pavia, and later rector of the university before devoting himself to political activities as the *podestà* of Cremona until 1859, when the city was annexed to the Kingdom of Italy. One of the more important questions highlighted by Baroli was the fact that Vattel had paid scant attention to the existence of a positive law of nations (the so-called voluntary law), thereby allowing an excessive pre-eminence of natural law over the positive law of states. Having entered the new century, it became easier for Baroli to underline how Vattel's work was a typically eighteenth-century product, and to demonstrate that new interpretations of the text were already imperative so that its significance and value could be circumscribed, interpreted and limited.

The upshot of all of these positions, from the German geographic area to the Italian one, was that the *Droit des gens* came to be considered a work irremediably tied to the past, to the context of the *ancien régime*. Thus it was no longer usable in the evolutionary sense, and it was not manageable in its possible interpretations unless accompanied by new commentaries and notes. This assertion was also made during the years of the Guerazzi-Romanelli trial by Pasquale Stanislao Mancini, future minister of public education and of foreign affairs in the Kingdom of Italy.[39] In his university lecture entitled *Della nazionalità come fondamento del diritto delle genti* (1851) Mancini called Vattel an ambiguous and superficial author whose work was at the very most useful as a resource from the viewpoint of historical and diplomatic documentation.[40] His criticisms of Vattel consisted of the proliferating idea that the *Droit des gens* was a mere synopsis of the works of Wolff, that its concept of 'nation' was ambiguous, and that it confused the ideas of 'state' and 'nation,' especially as regards the position of the individual and the citizen. In the background was the delicate problem of understanding whether the state, unlike the nation, was to be considered a consequence of natural law or an artificial construct.[41] Mancini was inclined towards the second idea, believing that the nation in particular was natural and therefore worthy of protection. This led him to unhesitatingly place Vattel in the group of "liberals," by which he meant the Enlightenment thinkers of the eighteenth century, and thus among the convinced supporters of the theories of the social contract. As Mancini wrote, the fact that Vattel had prefixed his reasoning with the "gospel of the social contract" had led him to mistakenly believe that only governments, rather than nations, were able to create legal ties. Consequently, the law of nations should be considered a natural law only of states and not of peoples.[42]

According to Mancini, Vattel had failed to reflect on the temporary nature of the link between the citizen and the state, overlooking the fact that this link could never be considered definitive. So much so, Mancini continued, that the answer to the question of whether a division or separation from the state was legitimate, and therefore if it was possible to question the sovereignty of the state itself, could only come from the application of the concept of nation as a right "to defend and govern themselves."[43] Hence the whole of the *Droit des gens* needed to be reread, Mancini continued, in the light of this principle, which placed the nation at the centre of the law of nations and international law. For example, the whole part relating to the signing, validity and interpretation of international treaties needed to be reconsidered, because these were to be accepted only if agreed for the good of the nation. Treaties established in the interests of the state or of a ruling party were worthless when they conflicted with the "good of the nation," as had happened in history with so many "odious" trade treaties, alliances and defence treaties. But Vattel, starting from one ambiguity or false premise or another, had ended up elaborating an explanation of the behaviour of the states, of the use of war and of international treaties, that seemed "horrendous and savage" and which had influenced too many authors after him.

THE DECLINE OF THE DROIT DES GENS IN NINETEENTH-CENTURY ITALY: VATTEL AND HIS "EXCESSES OF DEMOCRATIC SPIRIT"

The eclipse of the success of Vattel's work in Italy and the Mediterranean area occurred mainly in the 1850s and 1860s, exactly one century after its first publication. From that moment the *Droit des gens* was definitively consigned to literary history as a bibliographic document of the eighteenth century, as the outdated testimony of a world now gone, or at most as a polemical target with which to express the hardships and transformations of an era—like the mid-nineteenth century—which accompanied the formation of the great nation states and the repositioning of international power politics.

An interesting sign of how the reading of Vattel's work expressed this crisis and this transitional phase can be found in the near-forgotten work of a Neapolitan scholar, Terenzio Sacchi (1809–1865), who presented himself first of all as a state official and defender of the Bourbon monarchy, which was more isolated than ever on the international scene. Sacchi worked in the ministry of the interior, and was decidedly pro-monarchic and opposed to the 1848 revolution and the demands of the liberals for a new constitution in the Kingdom of the Two Sicilies. He gained prominence mainly through a long-distance argument with William Gladstone, who on returning from a visit to Naples in 1851 had published *Two letters to the Earl of Aberdeen, on the state prosecutions of the Neapolitan government*, which was addressed to the future Prime Minister, George Hamilton Gordon, and denounced the Bourbon police state.[44] Sacchi responded with a pamphlet in Italian intended more for internal debate than for international circulation, but he was soon convinced that behind Gladstone's criticism, which revealed the Tory party's position on the Italian and Mediterranean situation, there was a more articulated cultural and political strategy. This strategy revealed the principles of English power politics which could be recognised, according to Sacchi, by reading and interpreting the work of Vattel.

Sacchi's position is therefore interesting because it allows us once again to see how the use of the *Droit des gens* made it possible not only to identify the state's internal enemies, but also its external threats. Sacchi thus placed himself among those who in the eighteenth century had considered, instrumentally, Vattel to be an Anglophile, utilising the writings of Edmund Burke and others who presented him as a supporter of the dominion of the law and constitution of Britain and more generally of the beneficial role of Great Britain as a great power. This was a well-known notion according to which the *Droit des gens* helped to legitimise Great Britain's possession of colonial territories and its interest in a peaceful global commercial sphere, which would keep the world in order and in peace under British control.[45]

Sacchi was right about British interests in the Mediterranean, because by then for thirty years the English had been extending their range of influence in southern Italy, particularly Sicily and later the islands of the Aegean, not only through the development of a trade network but also through a strong political commitment to the spread of constitutional models based on the English example.[46] However, from the point of view of the impact on public opinion his answer would prove far less effective than he probably expected.

In fact, in the course of 1854 Sacchi began to publish a confutation or rewriting of the first book of the *Droit des gens*.[47] This was a book of eighty pages that contained a discussion and partial reworking of the first book of Vattel's work. Once again, then, the commentator's attention was concentrated on the part concerning the constitutional order of the state. Sacchi's work was published in monthly sections and distributed through a system of subscriptions. The copy preserved in the library of the Royal Palace in Naples has within it a small manifesto in which the author declared his intention to distribute the commentary, despite its being in Italian, in other parts of Europe.[48] He incorporated some of Pinheiro's notes in his text, but he was personally more penetrating in his criticism of Vattel, questioning the entire architecture of the *Droit des gens*. He underlined its dependency on Wolff, denounced its dangerousness for being full of "excesses of the democratic spirit" taken up by his imitators and admirers between the eighteenth and nineteenth centuries, and proposed another line of interpretation, that of François Guizot in the *Histoire de la civilisation en Europe*.[49] The reference to Guizot—defender of liberal monarchy and opponent of Orléanist constitutionalism—enabled Sacchi to link the interpretation of Vattel to the "actual state of the Nations" and to a notion of the social and moral progress of European civilisation that was natural and unstoppable but gradual, that had no need of revolutions. The reference to Guizot was no accident, because it evoked another great European power, France, whose financial interests in the Kingdom of the Two Sicilies, and in Naples in particular, were particularly extensive, making it an important rival of Great Britain for the control of the Mediterranean area.[50]

The result was that Vattel's book was reworked both in terms of form and of interpretation: of form because Sacchi changed the order of the paragraphs and devised a reading strategy completely different from that desired by the author. And it was different as regards interpretation because Sacchi, so as to bring it into a moderate political sphere compatible with the Bourbon culture and with the objectives of a constitutional monarchy, within the neo-absolutist framework, deprived the *Droit des gens* of those characteristics of generality and abstraction that had been its strengths a century earlier.

Proof that Sacchi's work was not, however, the personal initiative of a scholar passionate about the law of nations, but rather the reflection of a governmental position, if not indeed the end product of an assignment allocated directly by the government, comes from some little-known sources. In particular, a document of September 1855 signed by the intendant for the province of Bari, in Puglia, and sent to the region's mayors to explain that it had been the king himself who, the previous March, had issued instructions for Sacchi's commentary on Vattel to be sent out to all the communes and police headquarters, and for the subscriptions of the work to be promoted among scholars, jurists and the legal profession.[51]

As we have seen, the strategy pursued by Sacchi and the Bourbon government concerning the *Droit des gens* was intended to support an international position of the Kingdom of the Two Sicilies, which was then struggling to deal with British and French interests in the Mediterranean and the future political reorganisation of the Italian peninsula. Even so, Sacchi's work received a lukewarm welcome in the international press, although the *Revue critique de législation* and the *Morgen-Post* in Vienna wrote about it on 6 March 1857.[52] In reality the reasons for this failure probably resided in the fact that by the 1850s the *Droit des gens* no longer fired the political debate with enthusiasm, as it had once done.

NOTES

1. The scope of this cultural operation can be grasped through the work of Elisabetta Fiocchi Malaspina, *L'eterno ritorno del Droit des gens di emer de Vattel (secc. XVIII-XIX). L'impatto sulla cultura giuridica in prospettiva globale* (Frankfurt: Max Planck Institute for Legal History, 2017), 167 ff.
2. Robert Figge, *Georg Friedrich von Martens, sein Leben und sein Werke. Ein Beitrag zur Geschichte der Völkerrechtswissenschaft* (Breslau: Hill, 1914), 19.

3. Georg Friedrich Martens, *Primae lineae iuris gentium Europaerum practici in usum auditorum adumbratae* (Gottingue: Dieterich, 1785), 7, 138, 188, 242.

4. Georg Friedrich Martens, *Précis du droit des gens moderne de l'Europe fondé sulr les traités et l'usage. Pour servir d'introduction à un cours politique et économique* (Gottingue: Dieterich, 1801), 83, 97, 140, 160, 211, 283, 333, 392, 401, 425–430. The work was then constantly reprinted until the 1820s.

5. Georg Friedrich Martens, *Einleitung in das positive Europäische Völkerrecht auf Berträge und Herkommen gegründet* (Gottingue: Dieterich, 1796), 54, 260, 273, 300, 306, 339.

6. Martii Koskienniemi, "Into Positivism: Georg Friedrich von Martens (1756–1821) and Modern International Law," *Constellations* Volume 15, No 2 (2008), 190.

7. Wilhelm G. Grewe, *The Epochs of International Law*, transl. and rev. Michael Byers (Berlin-New York: De Gruyter, 2000), 291, 358, 410.

8. Philipp Mueller, "Archives and History: Towards a History of the 'Use of State Archives' in the 19th Century," *History of Human Science* 26 (2013): 27–49, https://journals.sagepub.com/doi/full/10.1177/0952695113502483

9. Henry Thomas Buckle, *History of Civilization in England* (London: Longmans Green and Co., 1873), vol. 2, 41, which refers to a comparison between Vattel, *Droit des gens*, and Ranke, *Geschichte der Päpste*.

10. Santi Di Bella, *Leopold von Ranke. Gli anni della formazione* (Soveria Mannelli: Rubettino, 2005), 61, 89.

11. Fiocchi Malaspina, *L'eterno ritorno*, 28, 111.

12. On this point, with a specific comparison between Ranke and Vattel, see Iver B. Neumann, "Status is Cultural: Durkheimian Poles and Weberian Russians Seek Great-Power Status," in *Status in World Politics*, ed. T. V. Paul, Deborah Welch Larson and William C. Wohlfort (Cambridge: Cambridge University Press, 2014), 89.

13. Theodore H. Von Laue, *Leopold von Ranke: The Formative Years* (Princeton: Princeton University Press, 1950), 34–38.

14. Fiocchi Malaspina, *L'eterno ritorno*, 51.

15. The editorial history is provided by Oke Manning, *Commentaries on the law of nations* (London: Sweet, 1839), 52–53.

16. Charles de Martens, "Avant-propos," in *Causes célèbres du droit des gens*, vol. 1 (Leipzig-Paris: Brockhaus-Ponthieu, 1827), XVI.

17. *Le droit des gens ou principes de la loi naturelle, appliqués à la conduite et aux affaires des Nations et des Souverains, revue et corrigée avec quelques remarques de l'editor, augmentée de quelques remarques nouvelles, et d'une Bibliographie choisie et systématique du droit des gens, par M. de Hoffmann*

précédée d'un Discours sur l'étude du droit de la nature et des gens, par sir James Mackintosh, [...] notes et table général analytique de l'ouvrage par Pinheiro Ferreira Silvestre (Paris: Aillaud, 1835–1838).

18. Sylvestre Pinheiro Ferreira, "Notes," in: Vattel (1838), Préliminaires, § 1, 1.

19. Benjamin Constant, *Ecrits Politiques -- Commentaire sur l'ouvrage de Filangieri (Oeuvres complètes, s. 1, t. 26)*, ed. Antonio Trampus and Kurt Kloocke (New York-Berlin: Walter de Gruyter, 2012).

20. On the relationship between the thought of Pinheiro and that of Constant, see José Estevan Pereira, *Silvestre Pinheiro: o su pensamento político* (Coimbra: Universidade de Coimbra, 1974), 125.

21. Fiocchi Malaspina, *L'eterno ritorno*, 182.

22. *Literarische Zeitung in Verbindung mit mehreren Gelehrten*, vol. 6 (1839), 75; Leopold August Warnkönig, *Rechtsphilosophie als Natulehre des Rechts* (Freiburg: Wagner, 1839), 434.

23. *Annali universali di statistica e economia pubblica, storia, viaggi e commercio*, vol. 69, no. 175 (1839): 8–10.

24. *Enciclopedia giuridica ad uso di lezioni* (Naples: Jovene, 1873), 473.

25. Ludovico Casanova, *Del diritto costituzionale lezioni* (Florence: Cammelli, 1875), 101.

26. Pasquale Fiore, *Trattato di diritto internazionale pubblico*, vol. 1 (Turin: Utet, 1887), 161.

27. John Norton Pomeroy, "To the State Board of Charities of the State of New York," in *Eight Annual Report of the State Board of Charities of the State of New York* (Albany: Weed, Parsons and Co., 1875), 144–148.

28. *Storia del processo politico di F.D. Guerrazzi ed altri imputati di perduellione corredata di documenti: Documenti* (Florence: G. Mariani 1851–1852). The second volume contains the trial documents.

29. *Storia del processo politico*, vol. 2.

30. Vattel, *Droit des gens*, book I, ch. XIX, § 220.

31. In relation to this Mari quoted Vattel, *Droit des gens*, book I, ch. XIX, § 203 and 204.

32. Vattel, *Droit des gens*, book II, ch. VII, § 79

33. *Storia del processo politico*, vol. 2, 59.

34. Vattel, *Droit des gens*, book II, ch. VII, § 93.

35. Alberto Carrera, "The citizen's right to leave his country: The concept of exile in Vattel's *Droit des gens*," in *The Legacy of Vattel's Droit des gens*, ed. Koen Stapelbroek and Antonio Trampus (Basingstoke: Palgrave Macmillan, 2019), 77–94.

36. Vattel, *Droit des gens*, book I, ch. XIX, § 229.

37. *Storia del processo politico*, vol. 2, 62.

38. Pietro Baroli, *Diritto naturale privato e pubblico* (Cremona: Feraboli, 1837), 88.

39. *Il pensiero liberale nell'età del Risorgimento*, ed. Valerio Castronovo (Rome: Istituto Poligrafico e Zecca dello Stato, 2001), 1237.

40. Pasquale Stanislao Mancini, *Della nazionalità come fondamento del dritto delle genti, prelazione al corso di diritto internazionale e marittimo pronunciata nella R. Università di Torino nel dì 22 gennaio 1851* (Turin: Botta, 1851), 19.

41. Arno Dal Ri Júnio and Chiara Sofia Mafica Biazi, "Debates a respeito do princípio de nacionalidade na doutrina italiana de dereito internacional da segunda metade do século XIX," *Revista da Facultade dereito UFMG* 70 (2017): 145–175.

42. Mancini, *Della nazionalità come fondamento del dritto delle genti*, 47.

43. Mancini, *Della nazionalità come fondamento del dritto delle genti*.

44. William Gladstone, *Two letters to the Earl of Aberdeen, on the state prosecutions of the Neapolitan government* (London: Murray, 1851), 6–8.

45. Paul W. Schroeder, "Did the Vienna Settlement Rest on a Balance of Power?" in *The American Historical Review* 97 (1992): 683–706; Richard Whatmore, *Against War and Britain, and France in the Eighteenth Century* (Yale: Yale University Press, 2012).

46. Rosa Maria Delli Quadri, *Il Mediterraneo delle costituzioni. Dalla Repubblica delle Sette Isole Unite agli Stati Uniti delle isole Ionie 1800–1817* (Milan: FrancoAngeli, 2017).

47. *Il diritto delle genti di E. de Vattel applicato allo stato attuale delle nazioni per Terenzio Sacchi* (Naples: Androsio, 1854).

48. Biblioteca del Palazzo reale di Napoli, no. XLI.124 (ex no 40.2.2).

49. On these elements see Fiocchi Malaspina, *L'eterno ritorno*, 242–243.

50. Gille Bertrand, *Les investissements français en Italie (1815–1914)* (Turin: ILTE, 1968).

51. The document was published in *Giornale dell'Intendenza della terra di Bari* 1 (Bari: Cannone, 1855), 198–199.

52. *Revue critique de legislation* (1856), 455–469; *Morgen-Post*, year 7, no. 64, (6 March 1857).

BIBLIOGRAPHY

Most of the quotations of Vattel's *Droit des gens* come from the widely available English-language edition of 2008, which includes an introduction by Béla Kapossy and Richard Whatmore (Indianapolis: Liberty Fund) and maintains the text and English title *The Law of Nations: Or, Principles of the Law of Nature Applied to the Conduct and Affairs of Nations and Sovereigns* of the 1797 London standard edition.

Contemporary translations of primary sources are listed under the names of their authors. Full manuscript sources are referenced only in the notes for reasons of space.

ARCHIVAL ABBREVIATIONS

BPRN Biblioteca del Palazzo Reale di Napoli, Naples, Italy

PRIMARY SOURCES

JOURNALS

Annali universali di statistica e economia pubblica, storia, viaggi e commercio, vol. 59, no. 175 (Milan: Società degli Annali Universali, 1839)
Giornale dell'Intendenza della terra di Bari, vol. 1 (Bari: Cannone, 1855)
Literarische Zeitung in Verbindung mit mehreren Gelehrten, vol. 6 (Berlin: Duncker & Humblot, 1839)
Morgen-Post Wien, year 7, no. 64, 6 March 1857
Revue critique de legislation et de jurisprudence, vol. 10, year 6 (Paris: Cotillon, 1856)

PRINTED BOOKS

Baroli, Pietro, *Diritto naturale privato e pubblico* (Cremona: Feraboli, 1837)
Casanova, Ludovico, *Del diritto costituzionale lezioni* (Florence: Cammelli, 1875)
Constant, Benjamin, *Écrits Politiques -- Commentaire sur l'ouvrage de Filangieri (Oeuvres complètes, s. 1, t. 26)*, eds. Antonio Trampus and Kurt Kloocke (New York-Berlin: Walter de Gruyter, 2012
Enciclopedia giuridica ad uso di lezioni (Naples: Jovene, 1873)
Fiore, Pasquale, *Trattato di diritto internazionale pubblico*, vol. 1 (Turin: Utet, 1887)
Gladstone, William, *Two letters to the Earl of Aberdeen, on the state prosecutions of the Neapolitan government* (London: Murray, 1851)
Mancini, Pasquale Stanislao, *Della nazionalità come fondamento del dritto delle genti, prelazione al corso di diritto internazionale e marittimo pronunciata nella R. Università di Torino nel dì 22 gennaio 1851* (Turin: Botta, 1851)
Manning, Oke, *Commentaries on the law of nations* (London: Sweet, 1839)
Martens, Charles de, "Avant-propos," in *Causes célèbres du droit des gens*, vol. 1 (Leipzig-Paris: Brockhaus-Ponthieu, 1827)
Martens, Georg Friedrich, *Einleitung in das positive Europäische Völkerrecht auf Berträge und Herkommen gegründet* (Gottingue: Dieterich, 1796)
———, *Précis du droit des gens moderne de l'Europe fondé sulr les traités et l'usage. Pour servir d'introduction à un cours politique et économique* (Gottingue: Dieterich, 1801)
———, *Primae lineae iuris gentium Europaerum practici in usum auditorum adumbratae* (Gottingue: Dieterich, 1785)

Pomeroy, John Norton, *To the State Board of Charities of the State of New York*, in *Eight Annual Report of the State Board of Charities of the State of New York* (Albany: Weed, Parsons and Co., 1875)

Sacchi, Terenzio, *Il diritto delle genti di E. de Vattel applicato allo stato attuale delle nazioni per Terenzio Sacchi* (Naples: Androsio, 1854)

Storia del processo politico di F.D. Guerrazzi ed altri imputati di perduellione corredata di documenti: Documenti (Florence: G. Mariani 1851–1852)

Vattel, Emer de, *Law of Nations: Or, Principles of the Law of Nature, Applied to The Conduct and Affairs of Nations and Sovereigns*, eds. Béla Kapossy and Richard Whatmore (Indianapolis: Liberty Fund 2008)

———, *Le droit des gens ou principes de la loi naturelle, appliqués à la conduite et aux affaires des Nations et des Souverains, revue et corrigée avec quelques remarques de l'editor, augmentée de quelques remarques nouvelles, et d'une Bibliographie choisie et systématique du droit des gens, par M. de Hoffmann précédée d'un Discours sur l'étude du droit de la nature et des gens, par sir James Mackintosh, [...] notes et table général analytique de l'ouvrage par Pinheiro Ferreira Silvestre* (Paris: Aillaud, 1835–1838)

Warnkönig, Leopold August, *Rechtsphilosophie als Natulehre des Rechts* (Freiburg: Wagner, 1839)

SECONDARY SOURCES

Bertrand, Gille, *Les investissements français en Italie (1815–1914)* (Turin: ILTE, 1968)

Buckle, Henry Thomas, *History of Civilization in England*, 2 vols., (London: Longmans Green and Co., 1873)

Carrera, Alberto, "The citizen's right to leave his country: The concept of exile in Vattel's *Droit des gens*," in *The Legacy of Vattel's Droit des gens*, ed. Koen Stapelbroek and Antonio Trampus (Cham: Palgrave Macmillan, 2019), 77–93

Castronovo, Valerio, ed., *Il pensiero liberale nell'età del Risorgimento* (Rome: Istituto Poligrafico e Zecca dello Stato, 2001)

Dal Ri Júnio, Arno and Mafica Biazi, Chiara, Sofia, "Debates a respeito do princípio de nacionalidade na doutrina italiana de dereito internacional da segunda metade do século XIX," *Revista da Facultade dereito UFMG 70* (2017), 145–175

Delli Quadri, Rosa Maria, *Il Mediterraneo delle costituzioni. Dalla Repubblica delle Sette Isole Unite agli Stati Uniti delle isole Ionie 1800–1817* (Milan: FrancoAngeli, 2017)

Di Bella, Santi, *Leopold von Ranke. Gli anni della formazione* (Soveria Mannelli: Rubettino, 2005)

Figge, Robert, *Georg Friedrich von Martens, sein Leben und sein Werke. Ein Beitrag zur Geschichte der Völkerrechtswissenschaft* (Breslau: Hill, 1914)

Fiocchi Malaspina, Elisabetta, *L'eterno ritorno del Droit des gens di Emer de Vattel (secc. XVIII–XIX). L'impatto sulla cultura giuridica in prospettiva globale* (Frankfurt: Max Planck Institute for Legal History, 2017)

Grewe, Wilhelm G., *The Epochs of International Law,* transl. and rev. by Michael Byers (Berlin-New York: De Gruyter, 2000)

Koskenniemi, Martii, "Into Positivism: Georg Friedrich von Martens (1756–1821) and Modern International Law," *Constellations* Volume 15, No 2 (2008), 189-207

Mueller, Philipp, "Archives and History: Towards a History of the 'Use of State Archives' in the 19th Century," *History of Human Science* 26 (2013), 27–49

Neumann, Iver B., "Status is Cultural: Durkheimian Poles and Weberian Russians Seek Great-Power Status," in *Status in World Politics,* eds. T. V. Paul, Deborah Welch Larson and William C. Wohlfort (Cambridge: Cambridge University Press, 2014), 85-114

Pereira, José Estevan, *Silvestre Pinheiro: o su pensamento político* (Coimbra: Universidade de Coimbra, 1974)

Schroeder, Paul W., "Did the Vienna Settlement Rest on a Balance of Power?" *The American Historical Review,* 97 (1992), 683–706

Von Laue, Theodore H., *Leopold von Ranke. The Formative Years* (Princeton: Princeton University Press, 1950)

Whatmore, Richard, *Against War and Britain, and France in the Eighteenth Century* (Yale: Yale University Press, 2012)

CHAPTER 12

Conclusion: Vattel's *Droit des gens* Between Good Government and Modern Democracy

Within the span of a century, from its publication in 1758 to the moment when two new great national states—Germany and Italy—appeared on the European political landscape, and thus at the time of the concomitant passing of the small state's role in the system of international balances, the political fortune of Emer de Vattel's *Droit des gens* became a kind of parable. Originally written for a practical purpose, the intention of its author was soon overlooked and the book became a political text in the era of the Atlantic Revolutions, then in the Restoration period it became the focal point of interpretative arguments, only to end up being regarded—in line with how eighteenth- and nineteenth-century culture represented it—as a straightforward legal manual, a simple synthesis of the natural law tradition, a witness to the eighteenth-century culture that became incompatible with the principles which inspired the formation of nation states and their positive law.

THE OTHER MODERNITY: VATTEL AND THE CONSTITUTION OF CÁDIZ (1812)

Though the *Droit des gens* was not a specific expression of the Enlightenment culture, but instead claimed continuity with the natural law tradition of previous centuries, it paradoxically became associated with many of the works by eighteenth-century authors that to varying degrees were held responsible for revolutionary disorder and over which a veil of silence

© The Author(s) 2020
A. Trampus, *Emer de Vattel and the Politics of Good Government*,
https://doi.org/10.1007/978-3-030-48024-0_12

gradually descended. This happened because the *Droit des gens*, with a level of awareness on the part of its author that still remains difficult to determine, took as central elements of its entire reasoning certain problems, such as the function of the constitutional state in the interstate system and the relationship between the exercise of good government and the wellbeing of citizens, destined to be at the centre of modern political debate, along with the concepts of nation, homeland and citizenship.

The last great period of the influence of Vattel's work over European constitutionalism was probably that which concerned the Constitution of Cádiz. In Spain, the *Droit des gens* had been condemned by the Inquisition in 1779, but it continued to circulate in private libraries and in universities by means of summaries and the teaching of professors.[1] In the late eighties and early nineties, Vattel's text thus became a key reference manual for teachers at the University of Salamanca, where many exponents of liberal culture had received their education. It is therefore no surprise that the *Plan de Educacion de la nobleza. Trabajado de Orden del Rey* recommended the reading of the *Droit des gens*, albeit in a censured version, referring to it as the most suitable instrument for understanding "precise and universal ideas" of use to any public office.

The timing and methods of this diffusion have not yet been studied in detail, but it is certain that Vattel was one of the authors through which consideration was given to the function of natural law in the development of the crisis of the *ancien régime* and in the transition from the idea of the political pact, understood from an abstract point of view as a fundamental law, to the idea of fundamental norms expressed through a positive law, namely the constitution. Consequently, as some scholars have been able to reconstruct, the *Droit des gens* certainly exerted influence on the cultural and political education of the generation active between the eighteenth and nineteenth centuries and on the activities of the early liberal intellectuals who wrote the text of the Constitution of Cádiz.[2]

An analysis of the preparatory work and the text of the constitution has given scholars a clearer understanding of the effect that the *Droit des gens* had on the development of the Iberian model of democracy. This was a model different to that of the Enlightenment and of the Atlantic revolutions, being based on the primacy of the Spanish nation and the Catholic nation, but at the same time firmly connected to a modern reinterpretation of the monarchy, an idea that rested on the recognition of individual rights and the primacy of the constitution. Bartolomé Clavero has written perceptively on the use of Vattel in the constitutional circles of Cádiz, demonstrating how the legislature, known as the Cortes, took the concept and meaning of the constitution as a written fundamental law from Vattel, as well as how it reinterpreted the concept of nation through him.[3]

Vattel had in fact explained quite clearly that the nation had an absolute right to create its own constitution, and to maintain and perfect it. At the same time, Spanish historiography has noted that entire articles of the constitution appear to have been written with Vattel's work to hand, particularly article 3, according to which sovereignty resided essentially in the nation, and thus because of this the exclusive right to establish its own fundamental laws belonged to it alone.[4]

A further contribution to understanding the importance of Vattel's work in the constitutional culture of Cádiz has been provided by José María Portillo Valdés who, studying the conception and practice of political representation in the Spanish kingdom after 1808, has identified in decree LXXII issued by the Cortes on 6 August 1811 another illuminating passage. This contains an explicit reference to the *Droit des gens*, made to explain and justify the incorporation of the jurisdictional authorities of the nation, in the belief that a free nation could be composed only of free men, not subject to any tie of dependence that excluded a free contract in the use of the sacred right to property.[5]

As the historian of law Bartolomé Clavero has observed, the Constitution of Cádiz was not exclusively Spanish, but was also American, African, a little Asian and a little Italian, because its model was propagated in all those places.[6] In this way, following the diffusion of the Spanish constitution, Vattel's work also found its way into the liberal constitutionalism of the nineteenth century and was brought into the Mediterranean area.

There no longer seems to be any doubt that the spread of the Constitution of Cádiz depended on the fact that it was a political model well suited to reconciling democratic aspirations with the monarchical tradition in liberal and anti-Napoleonic Europe.[7] In this sense Vattel's work, often reused only for certain parts or particular concepts, re-emerged during the Neapolitan revolution of 1821, as well as in the constitutional claims made in Piedmont, Modena, Bologna and Tuscany.

THE ECLIPSE OF THE DROIT DES GENS
IN NINETEENTH-CENTURY EUROPE

These are some of the reasons, identified in this research, which explain why within the history of Western, particularly European, culture the *Droit des gens* eventually became a victim of the polarisation of political discourse and language in the nineteenth century. Many ancient words began to sound off-key to the ears of the protagonists of the new century, and many central concepts of Vattel's reflections were applied and actualised with unexpected political consequences. In the *Droit des gens* the concepts of constitutional state and constitution had been used to assert the formal equality of all the countries in the interstate system, but they eventually found themselves squeezed between the demands of supporters of bottom-up sovereignty and those who preferred to remain connected to the idea of a concession by the prince. The theme of good government, to which the *Droit des gens* had devoted considerable space, describing it as a virtue and as a principle of republican inspiration, was transformed into a typical expression of the police state and neo-absolutist policies. Two concepts, those of the constitution and of good government, which to the ears of eighteenth-century men still spoke of programmes and desire for reform, innovation and institutional progress, in the Europe of the Restoration, especially in the Mediterranean region, became synonymous with conservatism and an antidemocratic spirit.

In the Piedmont of Victor Emmanuel I the good government of the *Droit des gens* became the magistracy (called by that very name: 'good government') placed in charge of the police, and the president of good government became, from 1814, the commander of the *carabinieri*. The law of 18 January 1815 asserted that the role of good government had to be "the maintenance of public tranquility and good order" in order to prevent crimes, in contrast to the justice system, which intervened to punish them. Ministries of good government and police were active in Modena, Tuscany and the Kingdom of the Two Sicilies during the Risorgimento era, while in Catholic Rome a Congregation of Good Government, in continuity with a similar magistracy of the sixteenth century, was until 1847 in charge of the economic and financial management of the communities of the papal state, exercising control over their budgets, land registers, population censuses and so on.[8]

It is therefore not possible to fully understand the history of the success and legacy of Emer de Vattel's *Droit des gens* within European states and internationally without taking into account these changed historical-geographical contexts and all of these strategies of appropriating and rejecting the text. This explains first of all why in the course of the nineteenth century an impressive cultural operation was undertaken to neutralise the political significance of the *Droit des gens* so as to ensure that it was no longer considered a key text for understanding and interpreting the politics of the states, but remained on library shelves to serve as a mere reference book, as a manual or as a compendium of information about a world that no longer existed. But the story of the shifting fortunes of the *Droit des gens* also helps us to understand better the reconfiguration of international balances, the change in the interstate dynamics of the nineteenth century, the eclipse of the role of the small state, and the subjugation of the ancient principle of the post-Westphalia balance to new power policies that would help redesign the geography of Europe.

This typically European phenomenon also allows us to understand why at the same time widely divergent ways of reading Vattel's work were being adopted in Europe, in North America and Latin America. In the cultural circles of the United States in particular, the trajectory of the *Droit des gens* was distinctive, thanks in part to the continuity of positive meanings of the concept of good government, associated both with the republican experience and a religious matrix.[9] This can be seen in the large number of discussions and reprints of the *Droit des gens* and in the fact that the book continued to enjoy practically uninterrupted success, such that between 1829 and 1854 one publisher alone, T. Johnson of Philadelphia, brought out eight editions of the book.[10]

Precisely for this reason, having returned from his journey to America, Alexis de Toqueville made Vattel one of the main reference points in his speeches to the chamber of deputies.[11] However, at that time the *Droit des gens*, like other classics of the European Enlightenment culture, was already available on the North American continent, and it was also enjoying a new lease of life in Latin America, where it accompanied the slow emergence of the nation states that moved away from the colonisation process by choosing the road of liberal constitutions. Vattel, like the Italian Filangieri and the Englishman Blackstone, would become an inspiration for constitutional reform in South America, which bore witness to a 'new renaissance' of the *Droit des gens*. Thus the Peruvian José Celedonio Urrea, the future leader of diplomatic negotiations with Uruguay and a defender of the culture of natural law against positivist European doctrines, was not wrong when he wrote that of all the eighteenth-century natural lawyers only Vattel had expounded his ideas in an immortal work destined to act as a long-running guide to solving international problems, compelling all subsequent writers to take due note of what he had to say.[12]

VATTEL'S THOUGHT TOWARDS MODERN DEMOCRACY

As I stated at the beginning of this book, Vattel's work has traditionally been considered a straightforward summary of the natural law tradition and a largely philosophical and legal text. As a result, research and historiography have often concentrated on the question of the originality of its contents, only to then stop in front of the impossibility of understanding how a text considered so unoriginal had experienced such great success and had exerted such a great influence on the history of Western culture.

In other words, a kind of intellectual short circuit ensued. The *Droit des gens*, as I have tried to show in this book, is a work that should be examined from unconventional observation points. When we say that Vattel was a transitional figure in the progression from the *ancien régime* to the modern age and that the *Droit des gens* was interpreted and had successes far beyond what the author (who died in 1767) had envisaged, this does not mean reducing the author's importance or the success story of the *Droit des gens* to a fashion for natural law.

Rather, the particularity of this work is due to the fact that the author knowingly and voluntarily put into place an act that was revolutionary from a cultural point of view, presenting the debate on the foundations of natural law and the law of nations to a non-academic readership and not only to an audience of jurists or philosophers, thereby placing this great store of concepts, definitions and ideas at the disposal of political circles. Thus it was these very circles in Europe and the Americas which, between the eighteenth and nineteenth centuries, used the *Droit des gens* and a significant part of the arguments contained within it (of good government, the constitution, the nation, the political virtue of small states, and the theory of rights) to orientate the intellectual debate towards the problem of the foundational characteristics of modern democracy.

The cases that I have presented, which focus on the Mediterranean and Italian area, and to a lesser degree on the German, Spanish and French-speaking states, testify to the many avenues—often forgotten by or unknown to the historiography of Vattel—through which all this happened. The *Droit des gens* thus became one of the cultural instruments used transnationally to bring about, or at least to call for, the actual transition from the *ancien régime* to the contemporary era. Many conservative circles within the Restoration were fully aware of this and, after failing to comprehensively present the *Droit des gens* as a defence of the old order, they preferred to launch an operation to neutralise the text politically and to present it as simply a legal technique manual. In this sense, speaking of the eclipse of the *Droit des gens* or of Vattel as a figure forgotten until the twentieth century does not mean telling the story of a work that disappeared from the history of cultural thought, but rather opens the way to the rediscovery of the many hidden paths that it has followed over the past two centuries.

NOTES

1. Francisco Tomás y Valiente, "Génesis de la Constitution de 1812," *Anuario de Historia del Derecho español* LXV (1995): 13–125.
2. José María Portillo Valdés, *Revolución de nación. Orígenes de la cultura constitucional en España* (Madrid: Boletín Oficial del Estado, 2000), 122–155; Francisco Tomás y Valiente, *Génesis de la Constitución de 1812. De muchas leyes fundamentales a una sola constitución* (Pamplona: Urgoiti, 2012), 34–35; Antonio Fernández García, *Las Cortes y la Constitución de Cádiz* (Madrid: Arco Libros, 2010), 47; Elisabetta Fiocchi Malaspina, "Vattel's Le droit des gens and the Constitution of Cádiz," in *Las Cortes de Cádiz y la Historia Parlamentaria*, ed. Diana Repeto García (Cádiz: Universidad de Cádiz, 2012), 33–40; Emiliano González Diez, "La monarquía constitucionalizada por la nación," in *Cádiz 1812. Origen del constitucionalismo español*, ed. Luis Palacio Bañuelos and Ignacio Ruiz Rodríguez (Madrid: Dyckinson S.L., 2013), 139.
3. Bartolomé Clavero, *Happy Constitution. Cultura y lengua constitucionales* (Madrid: Editorial Trotta, 1997), 168–171, 177, 180.
4. Portillo Valdés, *Revolución de nación*, 364–383.
5. Portillo Valdés, "Crisis e independencias: España y su monarquía," *Cuadernos dieciochistas* 8 (2007), 20–21.
6. Bartolomé Clavero, "Vocación catolica y advocación siciliana de la constitución española de 1812," in *Alle origini del costituzionalismo europeo*, ed. Andrea Romano (Messina: Accademia Peloritana dei Pericolanti, 1992), 11; Giorgio Spini, *Mito e realtà della Spagna nelle rivoluzioni italiane del 1820–1821* (Rome: Perrella, 1950); Antonino De Francesco, "La costituzione di Cadice nella cultura politica italiana del primo Ottocento," in *Rivoluzione e costituzione. Saggi sul democratismo politico nell'Italia napoleonica 1796–1821* (Naples: ESI, 1996), 142–146.

7. Andrea Romano, "Cadice come modello costituzionale per l'Europa liberale e antinapoleonica. Nota introduttiva," in *Costituzione politica della monarchia spagnola (1813)*, ed. Andrea Romano (Soveria Monnelli: Rubbettino 2000), XVII–LXXXV; Andrea Romano, "L'influenza della carta gaditana nel costituzionalismo italiano ed. europeo," in *La Constitucion de Cádiz de 1812: hacia los origenes del constitucionalismo iberoamericano y latino*, ed. Asdrúbal Aguliar Aranguren (Caracas: Universidad Catolica Andreas Bello, 2004), 351–373.

8. Filippo Sabetti, *The Search of Good Government: Understanding the Paradox of Italian Democracy* (Montreal: McGill-Queen's University Press, 2000), 26–51.

9. For example, see Charles Girdlestone, *The nature and value of good government and our Christian duties as good citizens and subjects especially in the present crisis* (Stourbridge: Mellard, 1848); George B. Loring, *Constitutional Freedom, the genius of our Government: An address delivered before the Columbian Society of Marblehead* (Boston: Boston Post, 1856), 19–20; Ezra Eastman Adams, *Government and Rebellion: A Sermon* (Philadelphia: Ashmed, 1861), 4–19.

10. There is a list in Elisabetta Fiocchi Malaspina, *L'eterno ritorno del Droit des gens di Emer de Vattel (secc. XVIII-XIX). L'impatto sulla cultura giuridica in prospettiva globale* (Frankfurt: Max Planck Institute for Legal History, 2017), 270–271.

11. David Clinton, *Tocqueville, Lieber, and Bagehot* (Basingstoke: Palgrave Macmillan, 2003), 30–31.

12. José Celedonio Urrea, *Principio de la lejislacion natural o filosofia del derecho* (Callao: Dañino, 1855), CVIII; Fabián Novak Talavera, *Las relaciones entre el Perú y la Francia (1827–2004)* (Lima: Pontificia Universidad Católica del Perú, 2005), 84–85.

BIBLIOGRAPHY

Most of the quotations of Vattel's *Droit des gens* come from the widely available English-language edition of 2008, which includes an introduction by Béla Kapossy and Richard Whatmore (Indianapolis: Liberty Fund) and maintains the text and English title *The Law of Nations: Or, Principles of the Law of Nature Applied to the Conduct and Affairs of Nations and Sovereigns* of the 1797 London standard edition.

Contemporary translations of primary sources are listed under the names of their authors. Full manuscript sources are referenced only in the notes for reasons of space.

PRIMARY SOURCES

PRINTED BOOKS

Adams, Ezra Eastman, *Government and Rebellion: A Sermon* (Philadelphia: Ashmed, 1861)

Girdlestone, Charles, *The nature and value of good government and our Christian duties as good citizens and subjects especially in the present crisis* (Stourbridge: Mellard, 1848)

Loring, George B., *Constitutional Freedom, the genius of our Government. An address delivered before the Columbian Society of Marblehead* (Boston: Boston Post, 1856)

Urrea, José Celedonio, *Principio de la lejislacion natural o filosofia del derecho* (Callao: Dañino, 1855)

Vattel, Emer de, *Law of Nations: Or, Principles of the Law of Nature, Applied to The Conduct and Affairs of Nations and Sovereigns*, eds. Béla Kapossy and Richard Whatmore (Indianapolis: Liberty Fund 2008)

SECONDARY SOURCES

Clavero, Bartolomé "Vocación católica y advocación siciliana de la constitución española de 1812," in *Alle origini del costituzionalismo europeo*, ed. Andrea Romano (Messina: Accademia Peloritana dei Pericolanti, 1992), 11–56

———, *Happy Constitution. Cultura y lengua constitucionales* (Madrid: Editorial Trotta, 1997)

Clinton, David, *Tocqueville, Lieber, and Bagehot* (Basingstoke: Palgrave Macmillan, 2003)

De Francesco, Antonino, "La costituzione di Cadice nella cultura politica italiana del primo Ottocento," in *Rivoluzione e costituzione. Saggi sul democratismo politico nell'Italia napoleonica 1796–1821* (Naples: ESI, 1996), 142–146

Fernández García, Antonio, *Las Cortes y la Constitución de Cádiz* (Madrid: Arco Libros, 2010)

Fiocchi Malaspina, Elisabetta, "Vattel's Le droit des gens and the Constitution of Cádiz," in *Las Cortes de Cádiz y la Historia Parlamentaria*, ed. Diana Repeto García (Cádiz: Universidad de Cádiz, 2012), 33–40

———, *L'eterno ritorno del Droit des gens di Emer de Vattel (secc. XVIII–XIX). L'impatto sulla cultura giuridica in prospettiva globale* (Frankfurt: Max Planck Institute for Legal History, 2017)

González Diez, Emiliano, "La monarquía constitucionalizada por la nación," in *Cádiz 1812. Origen del constitucionalismo español*, ed. Luis Palacio Bañuelos and Ignacio Ruiz Rodríguez (Madrid: Dyckinson S.L., 2013), 117–166

Novak Talavera, Fabián, *Las relaciones entre el Perú y la Francia (1827–2004)* (Lima: pontificia Universidad Católica del Perú, 2005)

Portillo Valdés, José María, *Revolución de nación. Orígenes de la cultura constitucional en España* (Madrid: Boletín Oficial del Estado, 2000)

———, "Crisis e independencias: España y su monarquía," *Cuadernos dieciochistas* 8 (2007), 19-35

Romano, Andrea, "Cadice come modello costituzionale per l'Europa liberale e antinapoleonica. Nota introduttiva," in *Costituzione politica della monarchia spagnola (1813)*, ed. Andrea Romano (Soveria Monnelli: Rubbettino 2000), XVII–LXXXV

———, "L'influenza della carta gaditana nel costituzionalismo italiano ed europeo," in *La Constitucion de Cádiz de 1812: hacia los origenes del constitucionalismo iberoamericano y latino*, ed. Asdrúbal Aguliar Aranguren (Caracas: Universidad Catolica Andreas Bello, 2004), 351–373

Sabetti, Filippo, *The Search of Good Government. Understanding the Paradox of Italian Democracy* (Montreal: McGill-Queen's University Press, 2000)

Spini, Giorgio, *Mito e realtà della Spagna nelle rivoluzioni italiane del 1820–1821* (Rome: Perrella, 1950)

Tomás y Valiente, Francisco, "Génesis de la Constitution de 1812," *Anuario de Historia del Derecho español* LXV (1995): 13–125

———, *Génesis de la Constitución de 1812. De muchas leyes fundamentales a una sola constitución* (Pamplona: Urgoiti, 2012)

BIBLIOGRAPHY

Despite the changes early modern books often underwent in translation, contemporary translations of primary sources are listed under the names of their authors. Full manuscript sources are referenced only in the notes for reasons of space.

ARCHIVAL ABBREVIATIONS

ASF	Archivio di Stato di Firenze, Florence, Italy
ASM	Archivio di Stato di Modena, Modena, Italy
ASN	Archivio di Stato di Napoli, Naples, Italy
ASS	Archivio di Stato di Siena, Siena, Italy
AST	Archivio di Stato di Torino, Turin, Italy
ASV	Archivio di Stato di Venezia, Venice, Italy
AV	Archivio Verri, Fondazione Raffaele Mattioli, Milan, Italy
BB	Bürgerbibliothek Bern, Berne, Switzerland
BP	Bibliothèque Patrimoniale de Bastia, Bastia, France
BPRN	Biblioteca del Palazzo Reale di Napoli, Naples, Italy
FQS	Fondazione Querini Stampalia, Venice, Italy
HHStA	Haus-, Hof- und Staatsarchiv Wien, Vienna, Austria
MCV	Museo Correr, Venezia, Venice, Italy
UA	Universiteitsbibliotheek Amsterdam, Amsterdam, The Netherlands

© The Author(s) 2020   229
A. Trampus, *Emer de Vattel and the Politics of Good Government*,
https://doi.org/10.1007/978-3-030-48024-0

PRIMARY SOURCES

Journals

Annali universali di statistica e economia pubblica, storia, viaggi e commercio, vol. 59, no. 175 (Milan: Società degli Annali Universali, 1839)

Giornale dell'Intendenza della terra di Bari, vol. 1 (Bari: Cannone, 1855)

Journal de commerce, vols. 2–3, april may (Bruxelles: Van den Berghen, 1759)

Journal politique [*Gazette des gazettes*], January 1–15, 1771 and February 1–15, 1771 (Bouillon: s.p., 1771).

Le Journal des Sçavans, June 7, 1688 (Paris: Académie des Inscriptions & belles-lettres, 1688)

Literarische Zeitung in Verbindung mit mehreren Gelehrten, vol. 6 (Berlin: Duncker & Humblot, 1839)

Morgen-Post Wien, year 7, no. 64, March 6 1857

Observateur hollandois ou quarantième lettre de M. van*** à M. H. de la Haye (La Haye: s.p., 1758)

Revue critique de legislation et de jurisprudence, vol. 10, year 6 (Paris: Cotillon, 1856)

Supplement du Journal des Sçavans, January 1707 (Paris: Delespine, 1707)

Printed Books

Adams, Ezra Eastman, *Government and Rebellion. A Sermon* (Philadelphia: Ashmed, 1861)

Alberti, Giovanni Giorgio degli, *Raccolta di tutto ciò che si è fin qui pubblicato in Livorno e altrove in morte di... Francesco I*, vol. 2 (Livorno: Strambi, 1765)

Azuni, Domenico Alberto, *Dizionario universale ragionato della giurisprudenza mercantile*, vol. 1 (Livorno: Masi, 1822)

———, *Principi del dritto marittimo dell'Europa*, 2 vols. (Trieste: Wage, Fleis & C., 1797)

Baroli, Pietro, *Diritto naturale privato e pubblico* (Cremona: Feraboli, 1837)

Beccaria, Cesare, *Dei delitti e delle pene. Novissima edizione di nuovo corretta, ed accresciuta coi commenti del Voltaire, confutazioni ed altri opuscoli interessanti di varj autori*, 2 vols. (Venice: Rinaldo Benvenuti, 1781)

———, *Dei delitti e delle pene*, critical edition by Gianni Francioni, in: *Opere*, vol. 1 (Milan: Mediobanca, 1984)

———, *Des délits et des peines*, ed. Philippe Audegean (Lyon: ENS, 2009)

Berni degli Antoni, Vincenzo, *Voto politico-legale per la città di Bologna* (Paris: n.p., 1831)

Bodin, Jean, *Le six livres de la République* (Paris: Puys, 1583)

Buonafede, Appiano, *Della restaurazione di ogni filosofia nei secoli 16., 17., 18.*, vol. 2 (Naples: Porcelli, 1788)

————, *Delle conquiste celebri esaminate col diritto natural delle genti* (Lucca: Riccomini, 1763)

Buondelmonti, Giuseppe Maria, *Ragionamento sul diritto della guerra giusta letto nell'Accademia della Crusca* (Florence: Bonducci, 1757)

Bynkershoek, Cornelis van, *De dominio maris dissertatio* [1702], in *Opera minora* (Leiden: van de Kerckhem, 1744)

Casanova, Giacomo, *Confutazione della storia del Governo Veneto d'Amelot de la Houssaie. Divisa in tré parti. Parte prima [-seconda]*, with *Supplimento all'opera intitolata Confutazione della storia del Governo Veneto d'Amelot de la Houssaie* (Amsterdam [Lugano]: Mortier [Agnelli], 1769)

————, *Examen des Études de la nature et Paul et Virginie de Bernardin de Saint-Pierre en 1788*, eds. Tom Vitelli and Marco Leeflang (Utrecht: L'Intermédiaire des Casanovistes, 1985)

————, *Exposition raisonné du différent, qui subsiste entre les deux Républiques de Venise et d'Hollande* ([Venise] : n.p., 1785a; it. translation *Esposizione ragionata della contestazione che sussiste tra le due Repubbliche di Venezia e d'Olanda* [Venezia]: n.p., 1785).

————, *Histoire de ma vie*, vol. III, eds. Gérard Lahouati and Marie-Françoise Luna (Paris: Gallimard, 2015)

————, *Le Messager de Thalie-Le Précis de ma Vie* (Paris: Fort, 1925)

————, *Pensieri libertini*, ed. Federico Di Trocchio (Milan: Rusconi, 1990)

————, *Supplément à l'exposition raisonné du différend qui subsiste entre les deux Républiques de Venise et d'Hollande* ([Venise], s.p. 1785b; it. translation *Supplimento alla esposizione ragionata della controversia che sussiste tra la Repubblica di Venezia e quella d'Olanda* ([Venice]: s.p., 1785)

Casanova, Ludovico, *Del diritto costituzionale lezioni* (Florence: Cammelli, 1875)

Chambrier, Jean-Pierre de, "Questions de droit des gens et Observations sur le Traité du Droit des Gens de M. de Vattel," in *Mémoires de l'Académie royale des Sciences et Belles-lettres depuis l'avénement de Fréderic Guillaume II au throne, MDCCLXXXVIII et MDCCLXXXIX* (Berlin: Decker, 1793), 436–492

Codignola, Ernesto, ed., *Carteggi di giansenisti liguri*, 3 vols. (Florence: Le Monnier, 1941–1942)

Constant, Benjamin, *Écrits Politiques – Commentaire sur l'ouvrage de Filangieri (Oeuvres complètes, s. 1, t. 26)*, eds. Antonio Trampus and Kurt Kloocke (New York-Berlin: Walter de Gruyter, 2012)

Correspondance inédite officielle et confidentielle de Napoleon Bonaparte-Venise (Paris: Panckoucke, 1819)

Covoni, Marco, *Orazione recitata nel solenne Capitolo de' Cavalieri di S. Stefano* (Florence: Allegrini, 1770)

De Simoni, Alberto, *Del furto e sua pena, trattato con alcune osservazioni generali in materia criminale* (Lugano: Agnelli, 1776)

Dragonetti, Giacinto, *Delle virtù e dei premi* (Modena: Montanari, 1768)

Enciclopedia giuridica ad uso di lezioni (Naples: Jovene, 1873)

Filangieri, Gaetano, *La scienza della legislazione. Edizione critica,* 7 vols., dir. Vincenzo Ferrone, eds. Antonio Trampus et al. (Venice: Edizioni della Laguna 2004)

Fiore, Pasquale, *Trattato di diritto internazionale pubblico,* vol. 1 (Turin: Utet, 1887)

Gaglio, Vincenzo, "Lettera al sig. Pepi sull'estrazione del feto vivente e morboso ne' parti pericolosi e difficili,"in *Opuscoli di autori siciliani,* vol. 19 (Palermo: Rupetti, 1778), 25–115

———, *Saggio sopra il diritto della natura e delle genti e della politica. Dell'avvocato Vincenzo Gaglio girgentino, accademico del Buon-Gusto* (Palermo: Valenza, 1759)

Galeani Napione, Gian Francesco, *Del modo di riordinare la Regia Università degli studi,* ed. Paola Bianchi (Turin: Deputazione Subalpina di Storia patria, 1993)

Galiani, Ferdinando, *De' doveri de' principi neutrali verso i principi guerreggianti e di questi verso i neutrali* ([Naples]: n.p., 1782)

Genovesi, Antonio, *Della diceosina, o sia della filosofia del giusto e dell'onesto,* ed. Niccolò Guasti (Venice: Centro di Studi sull'Illuminismo europeo-Edizioni della Laguna, 2008)

———, *Delle lezioni di commercio o sia di economia civile, con Elementi del commercio,* ed. Maria Luisa Perna (Naples: Istituto italiano per gli studi filosofici, 2005)

Girdlestone, Charles, *The nature and value of good government and our Christian duties as good citizens and subjects especially in the present crisis* (Stourbridge: Mellard, 1848)

Gladstone, William, *Two letters to the Earl of Aberdeen, on the state prosecutions of the Neapolitan government* (London: Murray, 1851)

Goudar, Ange, *Discours oratoire contenant l'éloge de son excellence monsieur le chevalier André Tron cy devant ministre extraordinaire en Holande et ambassadeur aux cours de Versailles et de Vienne* (Venice: Palese, 1773)

———, *Plan de reforme, propose aux cinq correcteurs de Venise actuellement en charge. avec un sermon evangelique pour elever la Republique dans la crainte de Dieu* (Amsterdam: n.p., 1775)

———, *The chinese spy* (London: Bladen, 1765)

Grisolia, Michelangelo, *Principj di diritto pubblico ovvero saggio sopra i libri del diritto della guerra e della pace* (Naples: Morelli, 1791)

Hervás y Panduro, Lorenzo, *Biblioteca jesuitico-española (1759–1793),* ed. Antonio Astorgano Arajo (Madrid: Libris, 2007)

Hume, David, *Essays Moral, Political, and Literary,* ed. Eugene Miller (Indianapolis: Liberty Fund, 1985)

Hume, David, *Of the Jealousy of Trade,* in *Essays and Treatises on Several Subjects, a new edition* (London: Millar, 1758)

Lampredi, Giovanni Maria, *De licentia in hostem liber singularis in quo Samuelis Cocceii sententia de infinita licentia in hostem exponitur et confutatur* (Florence: Excudebant Imperiales Typographi, 1761)

———, *Del commercio dei popoli neutrali in tempo di guerra* (Milan: n.p. 1788)

———, *Juris publici universalis sive juris naturae et gentium theoremata* (Livourne: n.p., 1776–1778)

Laugier, Marc Antoine, *Histoire de la République de Venise* (Venice: Carlo Palese and Gasparo Storti, 1767–1769)

Le Clerc, Jean, *Geschiedenissen der Vereenigde Nederlanden, sedert den aanvang van die Republyk tot op den Vrede van Utrecht in 't Jaar 1713* (Amsterdam: Zacharias Chatelain, 1738)

Lichtenberg, Georg Christoph, *Briefwechsel*, eds. Ulrich Joost and Albrecht Schöne, 3 vols. (Munich: Beck, 1983–1985)

Linguet, Nicolas-Henry-Samuel, *Annales politiques, civiles et littéraires du dix-huitième siècle* vol. 11 (London: Spilsbury and Snowhill, 1784)

———, ed., *Mélanges de politique et de littérature* (Bouillon: n.p., 1778)

Loring, George B., *Constitutional Freedom, the genius of our Government. An address delivered before the Columbian Society of Marblehead* (Boston: Boston Post, 1856)

Mancini, Pasquale Stanislao, *Della nazionalità come fondamento del dritto delle genti, prelazione al corso di diritto internazionale e marittimo pronunciata nella R. Università di Torino nel dì 22 gennaio 1851* (Turin: Botta, 1851)

Manning, Oke, *Commentaries on the law of nations* (London: Sweet, 1839)

Martens, Charles de, "Avant-propos," in *Causes célèbres du droit des gens*, vol. 1 (Leipzig-Paris: Brockhaus-Ponthieu, 1827)

Martens, Georg Friedrich, *Einleitung in das positive Europäische Völkerrecht auf Berträge und Herkommen gegründet* (Gottingue: Dieterich, 1796)

———, *Précis du droit des gens moderne de l'Europe fondé sur les traités et l'usage. Pour servir d'introduction à un cours politique et économique* (Gottingue: Dieterich, 1801)

———, *Primae lineae iuris gentium Europaerum practici in usum auditorum adumbratae* (Gottingue: Dieterich, 1785)

———, *Recueil des traités d'alliance, de paix, de trève etc.*, vols. 3 (Gottingue: Dieterich, 1818)

Mirabeau, Victor de Ruqueti, *Les devoirs: La scienza cioè i diritti e i doveri dell'uomo. Opera divisa in quattro parti che contengono 1. La vita naturale dell'uomo 2. La sua vita agricola 3. La sua vita sociale 4. La sua vita politica. Tradotta dalla prima edizione francese di Losanna dell'Anno 1773 da un accademico etrusco* (Florence: Cambiagi, 1774)

Newman, John, *The Character and Blessings of a Good Government: A Sermon Preach'd at Salters-Hall November 5th 1716* (London: Richard Ford, 1716)

Pagano, Francesco Mario, *Saggi politici. De' principii, progressi e decadenza della società (1791–92)*, eds. Luigi Firpo and Laura Salvetti Firpo (Naples: Vivarium, 1993)

Peter Leopold of Tuscany, *Relazioni sul governo della Toscana*, vols. 1–3, ed. A. Silvestrini (Florence: Olschki 1969–1974)

Pomeroy, John Norton, *To the State Board of Charities of the State of New York*, in *Eight Annual Report of the State Board of Charities of the State of New York* (Albany: Weed, Parsons and Co., 1875)

Progetto di costituzione della Repubblica napoletana presentato al governo provviso-rio dal comitato di legislazione, eds. Federica Morelli and Antonio Trampus, introduction by Anna Maria Rao (Venice: Centro di Studi sull'Illuminismo europeo, 2008)

Réal de Curban, Gaspard de, *La science du gouvernement, ouvrage de morale, de droit et de politique*, 5 vols. (Aix-La Chapelle [Amsterdam-Paris]: n.p., 1761–1765)

Ruzzini, Carlo, "Relatione del Congresso di Utrecht di miser Carlo Ruzzini, Kav. E Procurator, Ambasciatore estraordinario, plenipotenziario, 1713," in *Venetiaantsche Berichten Berichten over de Vereenigde Nederlanden 1600–1795*, ed. P. J. Blok (Gravenhage: Martinus Nijhoff, 1909)

Sacchi, Terenzio, *Il diritto delle genti di E. de Vattel applicato allo stato attuale delle nazioni per Terenzio Sacchi* (Naples: Androsio, 1854)

Sevigné, Madame de, *Choix de lettres françoises à l'usage des collèges & de tous ceux qui apprennent la langue Françoise* (Venice : Benvenuti, 1782; second edition Venice: Pasquali, 1788)

Sonnenfels, Joseph von, *Scienza del buon governo scritta dal Signor di Sonnenfels e recata dal Tedesco in italiano* (Milan: Giuseppe Galeazzi, 1784)

Storia del processo politico di F.D. Guerrazzi ed altri imputati di perduellione corre-data di documenti: Documenti (Florence: G. Mariani 1851–1852)

Urrea, José Celedonio, *Principio de la lejislacion natural o filosofia del derecho* (Callao: Dañino, 1855)

Vattel, Emer de, *Il diritto della natura e delle genti ovvero principii della legge naturale applicati alla condotta e agli affari delle nazioni e de' sovrani scritta nell'idioma francese dal Sig. di Vattel e recata nell'italiano da Lodovico Antonio Loschi* (Lyon [Venice]: s.p. [Vitto], 1781)

———, *Il diritto delle genti, ovvero Principii della legge naturale, applicati alla condotta e agli affari delle nazioni e de' sovrani. Opera scritta nell'idioma fran-cese dal sig. di Vattel e recata nell'italiano da Lodovico Antonio Loschi*, 3 vols. (Bologna: Fratelli Masi, 1804–1805)

———, *Law of nations* (Dublin: White, 1787)

———, *Law of Nations: Or, Principles of the Law of Nature, Applied to The Conduct and Affairs of Nations and Sovereigns*, eds. Béla Kapossy and Richard Whatmore (Indianapolis: Liberty Fund 2008)

——, *Le droit des gens ou principes de la loi naturelle appliqués à la conduite et aux affaires des nations et des souverains*. *Nouvelle édition augmentée* (Neuchâtel: Société Typographique, 1773)

——, *Le droit des gens ou principes de la loi naturelle appliquée à la conduite des affaires des nations et des souverains*. *Nouvelle edition* (Neuchâtel: De l'Imprimerie de la Société Typographique, 1777)

——, *Le droit des gens ou principes de la loi naturelle, appliqués à la conduite et aux affaires des Nations et des Souverains, revue et corrigée avec quelques remarques de l'editor, augmentée de quelques remarques nouvelles, et d'une Bibliographie choisie et systématique du droit des gens, par M. de Hoffmann précédée d'un Discours sur l'étude du droit de la nature et des gens, par sir James Mackintosh, [...] notes et table général analytique de l'ouvrage par Pinheiro Ferreira Silvestre* (Paris: Aillaud, 1835–1838)

——, *Mémoires politiques & militaires pour servir à l'histoire de notre tems*, vol. 2 (Frankfurt and Leipzig: 1760)

——, *Questions de droit naturel, et observations sur le Traité du droit de la nature de M. le baron de Wolf, par M. de Vattel* (Bern: Société typographique, 1762)

Vergani, Paolo Vergani, *Analisi ragionata del Congresso di Vienna*, 2 vols. (Genoa: Stamperia Pagano, 1818)

Verri, Alessandro, "Di alcuni sistemi del pubblico diritto", *Il Caffè* 32, 1766; reprinted in *Il Caffè 1764–1766*, eds. Gianni Francioni and Sergio Romagnoli, vol. 2 (Turin: Bollati Boringhieri, 1998), 736–739

Verri, Pietro, "Stato attuale del commercio di Milano," in *Opere*, vol. 2/2 *Scritti di economia, finanza e amministrazione*, eds. Giuseppe Bognetti, Angelo Moioli, Pier Luigi Porta and Giovanna Tonelli (Rome: Edizioni di Storia e Letteratura, 2006)

Warnkönig, Leopold August, *Rechtsphilosophie als Natulehre des Rechts* (Freiburg: Wagner, 1839)

Young, Edward, *Le lamentazioni, ossieno le Notti d'O. Y. coll'aggiunta di altre sue operette, libera traduzione di Ludovico Antonio Loschi* (Venice: Vitto, 1774)

Secondary Sources

Adorni Braccesi, Simonetta, and Ascheri, Mario (eds.), *Politica e cultura nelle Repubbliche italiane dal Medioevo all'età moderna. Firenze, Genova, Lucca, Siena, Venezia* (Rome: Istituto storico italiano per l'età moderna e contemporanea, 2001)

Aglietti, Marcella, *I governatori di Livorno dai Medici all'Unità d'Italia. Gli uomini, le istituzioni, la città* (Pisa: ETS, 2009)

Aiello, Raffele, *Preilluminismo giuridico e tentativi di codificazione nel regno di Napoli* (Naples: Jovene, 1965)

Aleksić, Branko, "Casanova et D'Alembert," *Recherches sur Diderot et l'Encyclopédie* 42 (2007), 83–94

Alimento, Antonella, "From Privilege to Equality: Commercial Treaties and the French Solutions to International Competition (1736–1770)," in *The Politics of Commercial Treaties in the Eighteenth Century. Balance of Power, Balance of Trade*, ed. Antonella Alimento and Koen Stapelbroek (Cham: Palgrave Macmillan, 2017), 243–266

———, "The French Reception of Vattel's Droit des gens: Politics and Publishing Strategies," in *The Legacy of Vattel's Droit des gens*, ed. Koen Stapelbroek and Antonio Trampus (Basingstoke: Palgrave Macmillan, 2019), 135–164

———, "Tra strategie editoriali e progettualità riformista: la circolazione in Francia de Le droit des gens di Emer de Vattel," *Rivista Storica Italiana* 129 (2017), 548–558

Andreoni, Anna and Demuru, Paola, *La facoltà politico-legale dell'università di Pavia nella Restaurazione (1815–1848)* (Milan: Cisalpino, 1999)

Arcidiacono, Bruno, "De la balance politique et des rapports avec les droits des gens: Vattel, la 'guerre pour l'équilibre' et le système européen," in *Vattel's International Law in a XXIst Century Perspective/Le droit International de Vattel vu du XXIe siècle*, ed. Vincent Chetail and Peter Haggenmacher (The Hague: Martinus Nijhoff, 2011), 77–100

Armellini, Serenella, *Libertà e legislazione. Il riformismo di Carlantonio Pilati* (Milan: Jaca Book, 1991)

Armitage, David, *Foundations of Modern Political Thought* (Cambridge: Cambridge University Press, 2013)

Aspremont, Jean d', *Formalism and the Sources of International Law. A Theory of the Ascertainment of Legal Rules* (Oxford: Oxford University Press, 2011)

Bachmann, Adrian, *Die preussiche Sukzession in Neuchâtel-Ein ständisches Verfahren um die Landesherrschaft im Spannungsfeld zwischen Recht und Utilitarismus (1694–1715)* (Zurich: Schulthess, 1993)

Balani, Donatella, *Il vicario tra città e stato. Ordine pubblico e annona nella Torino del Settecento* (Turin: Deputazione subalpina di storia patria, 1987)

Bar, Ludwig von, *A History of Continental Criminal Law* (Boston: Little, 1916)

Bayly, Christopher Alan and Biagini, Eugenio Federico, *Giuseppe Mazzini and the Globalization of Democratic Nationalism 1830–1920* (Oxford: Oxford University Press, 2008)

Bazzoli, Maurizio, *Il piccolo Stato nell'età moderna: Studi su un concetto della politica internazionale tra XVI e XVIII secolo* (Milan: Jaca Book, 1990)

———, *Stagioni e teorie della società internazionale* (Milan: LED, 2005)

Beaud, Oliviere, *La Puissance de l'État* (Paris: Presses Universitaires de France, 1994)

Beaulac, Stéphane, "Emer de Vattel and the Externalization of Sovereignty", *Journal of the History of International Law* 5 (2003), 237–292

Beguelin, Édouard, "En souvenir de Vattel (1714–1767)," in *Extrait du Recueil des travaux offerts par la Faculté de droit de l'Université de Neuchâtel à la Société suisse des Juristes à l'occasion de sa réunion à Neuchâtel 15–17 septembre 1929* (Neuchâtel: Université de Neuchâtel, 1929)

Belissa, Marc, *Fraternité universelle et intérêt national (1713–1795). Les cosmopolitiques du droit des gens* (Paris: Kimé, 1998)

Benzoni, Gino, "La città del Buon Governo: Venezia," in *Il Buono e il Cattivo Governo. Rappresentazioni nelle Arti dal Medioevo al Novecento*, ed. Giuseppe Pavanello (Venice: Marsilio, 2004), 93–108

Bertrand, Gille, *Les investissements français en Italie (1815–1914)* (Turin: ILTE, 1968)

Blair, Ann, "Authorial Strategies in Bodin," in *The Reception of Bodin*, ed. Howell A. Lloyd (Leiden-Boston: Brill, 2008), 137–156

Blösch, Eduard, *Johann Anton von Tillier*, in *Sammlung bernischer Biographien* vol. 2 (Bern: Francke, 1896), 542–547

Bobbio, Norberto, *The Future of Democracy: A Defence of the Rules of the Game* (Minneapolis: University of Minnesota Press, 1987)

Bois, Jean-Pierre, *De la paix des rois à l'ordre des empereurs 1714–1815* (Paris: Editions du Seuil, 2003)

Borgherini-Scarabellin, Maria, *Il Magistrato dei Cinque Savi alla Mercanzia dalla istituzione alla caduta della Repubblica: studio storico su documenti d'archivio* (Venice: Deputazione di storia patria per le Venezie, 1926)

Boucher, David, *Political Theories of International Relations:. From Thucydides to the Present* (Oxford: Oxford University Press, 1998)

Bozzola, Annibale, "Venezia e Savoia al congresso di Utrecht (1712–1713)", *Bollettino storico-bibliografico subalpino* XXXV.3–4 (1933), 30–39

———, *Giudizi e previsioni della diplomazia medicea sulla Casa di Savoia durante la guerra di successione di Spagna* (Turin: G. Bonis e Rossi, 1914)

Braida, Lodovica, *Il commercio delle idee. Editoria e circolazione del libro nella Torino del Settecento* (Florence: Olschki, 1995)

Bravetti, Patrizia and Granzotto, Orfea, *False date. Repertorio delle licenze di stampa veneziane con falso luogo di edizione (1740–1797)* (Florence: University Press, 2008)

Buckle, H. T., *History of Civilization in England*, 2 vols., (London: Longmans Green and Co., 1873)

Buonomo, Vincenzo, "Reciprocità, libertà religiosa e protezione dei diritti umani in ambito internazionale," in *Libertà religiosa e reciprocità*, ed. Jose Antonio Arana Mesa (Milan: Giuffrè, 2009), 119–148

Burkhard, Ernst, *Johann Anton von Tillier als Politiker* (Bern: Historischer Verein des Kantons Bern, 1963)

Cagnoli, Luigi, *Lodovico Antonio Loschi*, in *Notizie biografiche in continuazione della Biblioteca modenese del cavaliere abate Girolamo Tiraboschi*, vol. 5 (Reggio: Torregiani, 1837), 333–341

Calabria, Antonio, and Marino, John A., eds., *Good government in Spanish Naples* (Berlin-New York: Peter Lang, 1990)

Cancellier, Antonella and Grilli, Giuseppe, "La riflessione linguistica e traduttologica dei gesuiti in Italia: l'esempio di Masdeu," in *La presenza dei gesuiti iberici espulsi. Aspetti religiosi, politici,* culturali, eds. Ugo Baldini and Gian Paolo Brizzi (Bologna: Clueb, 2010), 577–586

Capra, Carlo, "The Italian States in Early Modern Period," in: *The Rise of the Fiscal State in Europe c. 1200–1815,* ed. Richard Bonney (Oxford: Oxford University Press, 1999), 417–439

———, *Giovanni Ristori da illuminista a funzionario (1755–1830)* (Florence: La Nuova Italia, 1968)

———, *I progressi della ragione. Vita di Pietro Verri* (Bologna: il Mulino 2002)

Carhart, Michael C., *The Science of Culture in Enlightenment Germany* (Cambridge, MA: Harvard University Press, 2007)

Carnino, Cecilia, *Lusso e benessere nell'Italia del Settecento* (Milan: FrancoAngeli, 2014)

Carpanetto, Dino, *Divisi dalla fede. Frontiere religiose, modelli politici, identità storiche nelle relazioni tra Torino e Ginevra (XVII–XVIII secolo)* (Turin: Utet, 2009)

Carrera, Alberto, "The citizen's right to leave his country: The concept of exile in Vattel's *Droit des gens*" in *The Legacy of Vattel's Droit des gens,* ed. Koen Stapelbroek and Antonio Trampus (Cham: Palgrave Macmillan, 2019), 77–93

Castronovo, Valerio, ed., *Il pensiero liberale nell'età del Risorgimento* (Rome: Istituto Poligrafico e Zecca dello Stato, 2001)

Cavanna, Adriano, *La codificazione penale in Italia. Le origini lombarde* (Milan: Giuffrè, 1987)

Chambrier, Frédéric-Alexandre de, *Histoire de Neuchâtel et Valangin jusqu'à l'avènement de la Maison de Prusse* (Neuchâtel: Attinger, 1840)

Chetail, Vincent, "Vattel et la sémantique du droit des gens: une tentative de reconstruction critique," in *Vattel's International Law in a XXIst Century Perspective/Le droit International de Vattel vu du XXIe siècle,* eds. Vincent Chetail and Peter Haggenmacher (The Hague: Martinus Nijhoff, 2011a), 385–434

———, and Haggenmacher, Peter, eds., *Vattel's International Law in a XXIst Century Perspective/Le droit International de Vattel vu du XXIe siècle* (The Hague: Martinus Nijhoff, 2011b)

Childs, James R., *Casanoviana. An Annotated World Bibliography* (Vienna: Nebehay, 1956)

Christov, Theodore, "Vattel's Rousseau: Jus Gentium and the Natural Liberty of States," in *Western Political Science Association 2011 Annual Meeting Paper.* Available at: https://ssrn.com/abstract=1766921, 167–169

Clavero, Bartolomé, "Vocación católica y advocación siciliana de la constitución española de 1812," in *Alle origini del costituzionalismo europeo*, ed. Andrea Romano (Messina: Accademia Peloritana dei Pericolanti, 1992), 11–56

———, *Happy Constitution. Cultura y lengua constitucionales* (Madrid: Editorial Trotta, 1997)

Clerici, Alberto, "Vattel in the Papal States: The Law of Nations and Anti-Prussian Propaganda in Italy at the Time of Seven Years' War," in *The Legacy of Vattel's Droit des gens*, ed. Koen Stapelbroek and Antonio Trampus (Cham: Palgrave Macmillan, 2019) 207–234

Clinton, David, *Tocqueville, Lieber, and Bagehot* (Basingstoke: Palgrave Macmillan, 2003)

Cochrane, Eric W., *Tradition and Enlightenment in the Tuscan Academies 1690–1809* (Rome: Edizioni di Storia e letteratura, 1961)

Comanducci, Paolo, *Settecento conservatore. Lampredi e il diritto naturale* (Milan: Giuffrè, 1981)

Conti, Vittorio, ed., *La ricezione di Grozio a Napoli nel Settecento* (Florence: Centro editoriale toscano, 2002)

Correnti, Santi, "Avvisaglie femministe nel Settecento siciliano," in *La Sicilia nel Settecento*, ed. Francesco Renda (Messina: Centro di Studi Umanistici, 1986), 129–132

Costa, Pietro, *Costituzione italiana: articolo 10* (Rome: Carocci, 2018)

———, "The Rule of Law: A Historical Introduction," in: *The Rule of Law: History, Theory and Criticism*, eds. Pietro Costa and Danilo Zoro (Dordrecht: Springer, 2007), 73–150

Crespo Solana, Ana, "Cooperation or Neutrality? How War Affects Business Strategies: The Case of Cadiz, Spain (1700–1720)", in *Neutres et neutralité dans l'espace atlantique durant le long XVIIIe siècle (1700–1820)*, ed. Éric Schnakenbourg (Bécherel: Les Perséides, 2015), 31–58

Crotti, Ilaria and Ricorda, Ricciarda, eds., *Gasparo Gozzi. Il lavoro di un intellettuale nel Settecento veneziano* (Padua: Editrice Antenore, 1989)

Cunningham, Allan, *Anglo-Ottoman Encounters in the Age of Revolution*, ed. by Edward Ingram, 2 vols. (London: Frank Cass, 1993)

Cusumano, Nicola, *Libri e cultura in Sicilia nel Settecento* (Palermo: New Digital Press, 2016)

D'Onofrio, Federico, "La 'nazione meglio polita': buon governo e costituzione economica della Cina alla scuola di Genovesi," in *Società e storia* 161 (2018), 471–497

Dal Ri Júnio, Arno and Mafica Biazi, Chiara, Sofia, "Debates a respeito do princípio de nacionalidade na doutrina italiana de dereito internacional da segunda metade do século XIX," *Revista da Facultade dereito UFMG* 70 (2017), 145–175

De Benedictis, Angela, "Contrattualismo e repubblicanesimo in una città d'antico regime: Bologna nello Stato della Chiesa," in *Materiali per una storia della cultura giuridica* 22 (1992), 269–299

———, "Nazione per diritto delle genti: Bologna città libera nello Stato della Chiesa," in *Nazioni d'Italia. Identità politiche e appartenenze regionali tra Settecento e Ottocento*, eds. Angela De Benedictis, Irene Fosi and Luca Mannori (Rome: Viella, 2012), 195–216

———, *Neither disobedients nor rebels: lawful resistence in early modern Italy* (Rome: Viella, 2018)

———, *Repubblica per contratto. Bologna: una città europea nello Stato della Chiesa* (Bologna: il Mulino, 1995)

De Francesco, Antonino, "La costituzione di Cadice nella cultura politica italiana del primo Ottocento," in *Rivoluzione e costituzione. Saggi sul democratismo politico nell'Italia napoleonica 1796–1821* (Naples: ESI, 1996), 142–146

Del Corno, Nicola, *Reazione*, in *Atlante culturale del Risorgimento. Lessico del linguaggio politico dal Settecento all'Unità*, eds. Alberto M. Banti, Antonio Chiavistelli, Luca Mannori and Marco Meriggi (Rome-Bari: Laterza, 2011), 163–167

Delia, Luigi, "The Enlightenment, Encyclopedism and the Natural Rights of Men: The Case of the Code of Humanity (1778)," in: *Thinking about the Enlightenment: Modernity and its Ramifications*, ed. Martin L. Davies (London-New York: Routledge, 2016), 69–85

Delli Quadri, Rosa, Maria, *Il Mediterraneo delle costituzioni. Dalla Repubblica delle Sette Isole Unite agli Stati Uniti delle isole Ionie 1800–1817* (Milan: FrancoAngeli, 2017)

Delogu, Giulia, "The Political Functions of Virtue in the Eighteenth-Century Italian Debate," *History of European Ideas* 43 (2017a), 889–913

———, *La poetica della virtù. Comunicazione politica e rappresentazione del potere in Italia tra Sette e Ottocento* (Milan: Misesis, 2017b)

Dessì, Rosa Maria, "Il bene comune nella comunicazione verbale e visiva. Indagini sugli affreschi del "Buon Governo"," in *Il Bene comune: forme di governo e gerarchie sociali nel basso medioevo* (XLVIII Convegno storico internazionale, Todi, 12–15 ottobre, 2011) (Spoleto: Centro italiano di studi sul l'alto medioevo: 2012), 89–130

———, "L'invention du 'Bon Gouvernement.' Pour une histoire des anachronismes dans les fresques d'Ambrogio Lorenzetti (XIVe–XXe siècle)," *Bibliothèque de l'Ecole de Chartes* 165 (2007), 129–180

Dezza, Ettore, "Dalle "scienze utili" alle "scientifiche professioni": la formazione universitaria di Giacomo Giovannetti," in *Saggi di storia del diritto penale moderno* (Milan: Giuffrè, 1992), 367–387

Di Bella, Santi, *Leopold von Ranke. Gli anni della formazione* (Soveria Mannelli: Rubettino, 2005)

Di Carlo, Gaetano, *Vincenzo Gaglio, Un intellettuale nella Girgenti del Settecento tra giurisprudenza e archeologia* (doctoral thesis, University of Pisa, 2015–2016),

Di Giovanni, Vincenzo, "Le origini delle accademie dei Riaccesi e del Buon Gusto (1568, 1622, 1718)" in *L'accademia nazionale di scienze lettere e arti di Palermo, 1718–1984: note storiche pubblicate in onore dei partecipanti alla 59. sessione della Union Academique Internationale riunita a Palermo dal 2 all'8 giugno 1985* (Palermo: Accademia di Scienze Lettere e Arti di Palermo, 1985), 19–54

Di Renzo Villata, Gigliola, *Formare il giurista: esperienze nell'area lombarda tra Sette e Ottocento* (Milan: Giuffrè, 2004)

Di Rienzo, Eugenio, *Decadenza e caduta del cosmopolitismo: Francia/Europa, 1792–1848. Note per una ricerca*, in *L'idea di cosmopolitismo: circolazione e metamorfosi*, ed. Lorenzo Bianchi (Naples: Liguori, 2002), 419–458

Di Simone, Maria Rosa, *Legislazione e riforme nel Trentino del Settecento: Francesco Vigilio Barbacovi tra assolutismo e Illuminismo* (Bologna: il Mulino, 1992)

Diaz, Furio, *Buondelmonti Giuseppe Maria*, in *Dizionario biografico degli italiani*, vol. 15 (Rome: Istituto della Enciclopedia Italiana, 1972), 212–215

———, *Francesco Maria Gianni. Dalla burocrazia alla politica sotto Pietro Leopoldo di Toscana* (Milan-Naples: Ricciardi, 1966)

Dioguardi, Gianfranco, *Ange Goudar contro l'Ancien Régime* (Palermo: Sellerio, 1988)

Emery, Ted, "Casanova's Coffeehouse: Sociability, Social Class, and the Well-bred Reader in Histoire de ma vie," in *The Thinking Space. The Café as a Cultural Institution in Paris, Italy and Vienna*, eds. Leona Rittner, W. Scott Haine and Jeffrey H. Jackson (London-New York: Routledge, 2016), 169–184

Faber, Eva, "Die Ehe der Gräfin Giustiniana Rosenberg-Wynne (1737–1791)," in *Adel im «langen» 18. Jahrhundert*, eds. G. Haug-Moritz et al. (Vienna: Verlag Österreichische Akademie der Wissenschaften, 2009), 289–310.

Fabrizio Lomonaco, *A partire da Giambattista Vico. Filosofia, diritto e letteratura nella Napoli del secondo Settecento* (Rome: Edizioni di Storia e Letteratura, 2010a)

Feller, Richard and Bonjour, Edgar, *Geschichtsschreibung der Schweiz*, 2 vols. (Basel: Schwabe, 1979)

Feola, Raffaele, *Dall'Illuminismo alla Restaurazione. Donato Tommasi e la legislazione delle Sicilie* (Naples: Jovene, 1982)

Fernández García, Antonio, *Las Cortes y la Constitución de Cádiz* (Madrid: Arco Libros, 2010)

Ferrari, Stefano, and Romagnani, Gian Paolo, eds., *Carlantonio Pilati: Un intellettuale trentino nell'Europa dei lumi* (Milan: FrancoAngeli, 2005)

Ferrone, Vincenzo, *Storia dei diritti dell'uomo. L'Illuminismo e la costruzione del linguaggio poitico dei moderni* (Rome-Bari: Laterza, 2014)

———, *The Politics of Enlightenment: Constitutionalism, Republicanism and the Rights of Men* (London-New York: Anthem Press, 2012)

Figge, Robert, *Georg Friedrich von Martens, sein Leben und sein Werke. Ein Beitrag zur Geschichte der Völkerrrechtswissenschaft* (Breslau: Hill, 1914)

Fiocchi Malaspina, Elisabetta, "Vattel's *Le droit des gens* and the Constitution of Cádiz," in *Las Cortes de Cádiz y la Historia Parlamentaria*, ed. Diana Repeto García (Cádiz: Universidad de Cádiz, 2012), 33–40

———, *L'eterno ritorno del Droit des gens di Emer de Vattel (secc. XVIII–XIX). L'impatto sulla cultura giuridica in prospettiva globale* (Frankfurt: Max Planck Institute for Legal History, 2017)

Francioni, Gianni, "Nota al testo," in: Cesare Beccaria, *Dei delitti e delle pene*, critical edition by Gianni Francioni, *Opere*, vol. 1 (Milan: Mediobanca, 1984), 215–368

Friedrich, Manfred, *Geschichte der deutschen Staatsrechtswissenschaft* (Berlin: Duncker und Humblot, 1997)

Frigo, Daniela, *Principi ambasciatori e jus gentium. L'amministrazione della politica estera nel Piemonte del Settecento* (Rome: Bulzoni, 1991)

Frosini, Tommaso Edoardo, "Is good government a myth ?," in Sabetti, Filippo, (ed.), *Alla ricerca del buon governo in Italia* (Manduria: Lacaita, 2004), 5–10

Gabba, Emilio, and Schiavone, Aldo (eds.), *Polis e piccolo Stato tra riflessione antica e pensiero moderno (atti della giornate di studio 21–22 febbraio 1997)* (Florence: New Press, 1997).

Gaeta, Rinaldo, *Carlo Antonio Pilati. Dalle esperienze culturali europee al riformismo trentino (1760–1802)* (Venice: Deputazione di Storia Patria per le Venezie, 1995)

García Mora, Manuel R., *International Law and Asylum as a Human Right* (Washington: Public Affairs Press, 1956)

Gaudin, Jean-Pierre, "Modern governance, yesterday and today: some clarifications to be gained from French government policies," *International Social Science Journal* 50 (155) (1998), 47–56

Giani, Marco, "Paolo Paruta: il lessico della politica" (doctoral thesis, Ca' Foscari University, 2011)

Giarrizzo, Giuseppe, "Appunti per la storia culturale della Sicilia settecentesca," in *Rivista Storica Italiana* 79 (1967), 573–627

Glanville, Luke, "Historical Thinking about Human Protection. Insights from Vattel," in *Routledge Handbook of Ethics and International Relations*, eds. Brent J. Steele and Eric A. Heinze (London: Routledge, 2018) 308–317

Godoli, Ezio, *Trieste* (Rome-Bari: Laterza, 1984)

González Diez, Emiliano, "La monarquía constitucionalizada por la nación," in *Cádiz 1812. Origen del constitucionalismo español*, ed. Luis Palacio Bañuelos and Ignacio Ruiz Rodríguez (Madrid: Dyckinson S.L., 2013), 117–166

Good, Christoph, *Emer de Vattel (1714–1767). Naturrechtliche Ansätze einer Menschenrechtsidee und des humanitäre Völkerrecht im Zeitalter der Aufklärung* (Zurich: Dike Verlag AG, 2011).

Grahl-Madsen, Atle, *The Status of Refugees in International Law: Asylum, entry and sojourn* 2 vols. (Leiden: Sijthoff, 1966)

Grandpierre, Louis, *Histoire du Canton de Neuchâtel sous le Roi de Prusse 1707–1848* (Leipzig: Grandpierre, 1805)

Gresky, Wolfgang, "Der Reichsgraf Johann Ludwig von Wallmoden-Gimborn und sein Schlösschen im Georgengarten'" *Hannoverscher Geschichtsblätter* 36 (1982), 252–279

Grewe, Wilhelm G., *The Epochs of International Law*, transl. and rev. by Michael Byers (Berlin-New York: De Gruyter, 2000)

Guasti, Niccolò, *L'esilio italiano dei gesuiti spagnoli. Identità, controllo sociale e pratiche culturali (1767–1798)* (Rome: Edizioni di Storia e Letteratura, 2006)

———, "Un caso editoriale: la Diceosina di Antonio Genovesi", in Genovesi, Antonio, *Della diceosina, o sia della filosofia del giusto e dell'onesto*, ed. Niccolò Guasti (Venice: Centro di Studi sull'Illuminismo europeo-Edizioni della Laguna, 2008), xi–lxxi

Haitsma Mulier, E. O. G., "De affaire Zanovich. Amsterdams-Venetiaansche betrekkingen aan het einde van de achttiende eeuw," *Amstelodamum* 72 (1980), 85–119

Hauc, Jean-Claude, *Ange Goudar: Un aventurier des Lumières* (Paris: Honoré Champion, 2004)

Heineman, Robert, *Authority and Liberal Tradition from Hobbes to Rorty* (New Brunswick-London: Transaction Publishers, 1994)

Hont, Istvan, *Jealousy of Trade: International Competition and the Nation-State in Historical Perspective* (Cambridge, MA: Harvard University Press, 2005)

Hunt, Lynn, *Inventing Human Rights: A History* (New York: Norton & Co., 2008)

Hurrell, Andrew, *Vattel: Pluralism and its Limits*, in *Classical Theories of International Relations*, eds. Ian Clark and Iver Neumann (Houndmills-New York: Palgrave, 2001), 233–255

Ingerbritsen, Christine, ed., *Small states in International relations* (Reykjavik: University of Iceland Press, 2006)

Isabella, Maurizio, *Risorgimento in Exile: Italian Emigrés and the Liberal International in the Post-Napoleonic Era* (Oxford: Oxford University Press, 2009)

Jacona, Erminio and Turrini, Patrizia, *Le carte Brichieri Colombi. Inventario analitico* (Rome: Ministero per i beni e le attività culturali-Direzione generale per gli archivi, 2003)

Johnston, Douglas M., *Historical Foundations of World Order: The Tower and the Arena* (Leiden-Boston: Martinus Nijhoff, 2008)

Jouannet, Emmanuelle, "Les dualismes du Droit des gens," in *Vattel's International Law from a XXIst Century Perspective*, eds. Vincent Chetail and Peter Haggenmacher (Leiden: Brill, 2011), 133–150

Kapossy, Béla, and Wathmore, Richard, "Introduction," to *Law of Nations: Or, Principles of the Law of Nature, Applied to The Conduct and Affairs of Nations and Sovereigns*, eds. Béla Kapossy and Richard Whatmore (Indianapolis: Liberty Fund 2008), ix–xx

Keping, Yu, "Governance and Good Governance: A Framework for Politica Analysis," *Fudan Journal of the Humanities and Social Science* 11 (1) (March 2018), 1–8.

Kontler, László, "Polizey and Patriotism: Joseph von Sonnenfels and the Legitimacy of Enlightened Monarchy in the gaze of eighteenth-century State Sciences," in *Monarchism and Absolutism in Early Modern Europe*, eds. Cesare Cuttica and Glenn Burgess (London and New York: Routledge, 2012), 75–90

Kosary, Domokos, *Les "petits Etats" faceaux changements culturels, politiques et économiques de 1750 à 1914* (Lausanne: HU Jost, 1985)

Koskenniemi, Martii, ""International Community" from Dante to Vattel," in *Vattel's International Law in a XXIst Century Perspective/Le droit International de Vattel vu du XXIe siècle*, eds. Vincent Chetail and Peter Haggenmacher (The Hague: Martinus Nijhoff, 2011), 51–75.

———, "Into Positivism: Georg Friedrich von Martens (1756–1821) and Modern International Law," *Constellations* Volume 15, No 2 (2008), 189–207

Kratochwil, Friedrich, "On the notion of "interest" in international relations," in *International organizations* 36 (1982), 1–30

Langewiesche, Dieter, ed., *Kleinstaaten in Europa: Symposium am Liechtenstein-Institut zum Jubiläum 200 Jahre Souveränität Fürstentum Liechtenstein 1806–2006* (Vaduz: Verlag der Liechtensteinischen Akademischen Gesellschaft, 2007)

Lapradelle, Albert Geouffre de, "Introduction" to Emer de Vattel, Le Droit des Gens ou principes de la loi naturelle appliqués à la conduite et aux affaires des nations et des souverains (Washington: Carnegie Institution, 1916), iii–lix

Laursen, John Christian Laursen, Blom, Hans and Simonutti, Luisa (eds.), *Monarchism in the Age of Enlightenment. Liberty, Patriotism and Common Good*, (Toronto: University of Toronto Press, 2007)

Lebeau, Christine, "Negotiating a Trade Treaty in the Imperial Context: The Habsburg Moarchy in the Eighteent Century," in *The Politics of Commercial Treaties in the Eighteenth Century. Balance of Power, Balance of Trade*, eds. Antonella Alimento and Koen Stapelbroek (Basingstoke: Palgrave Macmillan, 2017), 349–369

Lehner, Ulrich L., *The Catholic Enlightement: The Forgotten History of a Global Movement* (Oxford: Oxford University Press, 2016)

Leonhard, Jörn, *Liberalismus: zur historischen Semantik eines europäischen Deutungsmuster* (Munich: R. Oldenbourg Verlag, 2001)

Lippert, Woldermar, *Kaiserin Maria Theresia und Kurfuerstin Maria Antonia von Sachsen: Briefwechsel, 1747–1772* (Leipzig: Teubner, 1903)

Loescher, Gil, ed. *Refugees and the Asylum Dilemma in the West*, (Pennsylvania: Pennsylvania State University Press, 1992)

Lomonaco, Fabrizio, *A partire da Giambattista Vico. Filosofia, diritto e letteratura nella Napoli del secondo Settecento* (Rome: Edizioni di Storia e Letteratura, 2010b)

Luna, Marie-Françoise, "Giacomo Casanova de Seingalt (1725–1798)," in *Dictionnaire des journalistes (1600–1789)*, eds. Anne-Marie Mercier-Faivre and Denis Reinaud (2010), (http://dictionnaire-journalistes.gazettes18e.fr/journaliste/143-giacomo-casanova-de-seingalt).

Maass, Matthias, *Small states in world politics: The story of small states survival 1648–2016* (Manchester: Manchester University Press, 2017)

Mafrici, Mirella, *Il mezzogiorno d'Italia e il mare: problemi difensivi nel Settecento*, http://www.storiamediterranea.it/public/md1_dir/b703.pdf, 657–658.

Manfrin, Mauro, "La famiglia von Martens alla Mira vecchia," *Rive* 8 (2011), 72–79

Mannori, Luca, *Il sovrano tutore. Pluralismo istituzionale e accentramento amministrativo nel principato dei Medici (sec. XVI–XVIII)* (Milan: Giuffrè, 1994)

———, and Sordi, Bernardo, *Storia del diritto amministrativo* (Rome-Bari: Laterza, 2003)

Mars, Francis L., "Ange Goudar, cet inconnu (1708–1791)", *Casanova Gleanings* IX (1966), 43–44

Mastromartino, Fabrizio, *Il diritto d'asilo. Teoria e storia di un istituto giuridico controverso* (Turin: Giappichelli, 2012)

Mazzone, Umberto, *'El buon governo:' un progetto di riforma generale nella Firenze savonaroliana* (Florence: Olschki, 1978)

Mena, Fabrizio, *Stamperie ai margini dell'Italia. Editori e librai nella Svizzera italiana 1746–1848* (Bellinzona: Casagrande, 2003)

Miglio, Gianfranco, *La controversia sui limiti del commercio neutrale fra Giovanni Maria Lampredi e Ferdinando Galiani* (Milan: Ispi, 1942)

Monti, Gennaro Maria, *La dottrina dell'abate F. Galiani sulla neutralità e l'adesione di Ferdinando IV alla Lega dei Neutri* (Milan: Ispi, 1942)

Monticelli, Chiara Lucrezio, *La polizia del papa: istituzioni di controllo sociale a Roma nella prima metà dell'Ottocento* (Soveria Mannelli: Rubbettino, 2012)

Mortier, Roland, *Le "prince d'Albanie". Un aventurier au siècle des Lumières* (Paris: Honoré Champion, 2000)

Moschetti, Carlo Maria, ed., *Il Codice marittimo del 1781 di Michele de Jorio per il regno di Napoli*, 2 vols. (Naples: Giannini 1979)

Mueller, Philipp, "Archives and History: Towards a History of the 'Use of State Archives' in the 19th Century," *History of Human Science* 26 (2013), 27–49

Musi, Aurelio, "La nazione napoletana prima della nazione italiana," in *Nazioni d'Italia. Identità politiche e appartenenze regionali tra Settecento e Ottocento*, eds. Angela De Benedictis, Irene Fosi and Luca Mannori (Rome: Viella, 2012), 75–89

Nakhimovsky, Isaac, "Carl Schmitt's Vattel and the *Law of Nations* between Enlightenment and Revolution," *Grotiana* 31 (2010), 141–164

———, "Vattel's Theory of the International Order: Commerce and Balance of Power in the Law of Nations," *History of European Ideas* 33 (2007), 157–173

Neff, Stephen C., "Peace and prosperity: commercial aspects of peacemaking," *Peace treaties and International Law in European History. From the Late Middle Ages to World War One*, ed. Randall Lesaffer (Cambridge: Cambridge University Press, 2004), 365–381

———., *The rights and duties of neutrals: A general history* (Manchester: Manchester University Press, 2000)

Neumann, Iver B., "Status is Cultural: Durkheimian Poles and Weberian Russians Seek Great-Power Status," in *Status in World Politics*, eds. T. V. Paul, Deborah Welch Larson and William C. Wohlfort (Cambridge: Cambridge University Press, 2014), 85–114

Novak Talavera, Fabián, *Las relaciones entre el Perú y la Francia (1827–2004)* (Lima: pontificia Universidad Católica del Perú, 2005)

Onnekink, David, *The treaty of Utrecht 1713*, in *Peace was made here: the treaties of Utrecht, Rastatt and Baden 1713–1714*, eds. Renger de Bruin and Maarten Brinkman (Petersberg: Im Hof, 2013), 60–69

Onuf, Nicholas Grenwood. "Civitas Maxima: Wolff, Vattel and the fate of Republicanism," *American Journal of International Law* 88 (1994), 287–296

Palermo, Antonio, *Vincenzo Gaglio e il Rinnovamento siciliano* (Agrigento: Siculgrafica, 2017)

Pasta, Renato, "Beccaria tra giuristi e filosofi: aspetti della sua fortuna in Toscana e nell'Italia centrosettentrionale," in *Cesare Beccaria tra Milano e l'Europa*, eds. Sergio Romagnoli and Gian Domenico Pisapia (Milan-Rome: Cariplo-Laterza, 1990), 512–533

Patalano, Rosario, and Reiner, Sophus A., *Introduction*, to *Antonio Serra and the Economics of Good Government* (London: Palgrave Macmillan, 2016)

Pattison, Robert, *The Great Dissent: John Henry Newman and the Liberal Heresy* (Oxford: Oxford University Press, 1991)

Pereira, José Estevan, *Silvestre Pinheiro: o su pensamento político* (Coimbra: Universidade de Coimbra, 1974)

Pilati, Giuseppe, *Cenni su la vita e su le opere di Carlo Antonio Pilati stesi per la prima volta coll'aiuto di documenti da un Trentino* (Rovereto: Sottochiesa, 1874)

Pizzetti, Silvia Maria, "La costruzione della pace e di una società internazionale nell'Europa moderna fra jus gentium e cosmopolitismo (secoli XVII–XVIII)," in *Con la ragione e col cuore. Studi dedicati a Carlo Capra*, eds. Stefano Levati and Marco Meriggi (Milan: FrancoAngeli, 2008), 209–241

Pocock, John G. A., *Barbarism and Religion*, vol. 1, *The Enlightenments of Edward Gibbon, 1737–1764* (Cambridge: Cambridge University Press, 1999), vol. 2, *Narratives of Civil Government* (Cambridge: Cambridge University Press, 1999)

Poncelet, Christian, Büttiker Rolf, and Busino, Giovanni, eds., *Genève et la Suisse dans la pensée politique: actes du colloque de Genève (14–15 septembre 2006* (Marseille: Presses Universitaires d'Aix-Marseille, 2007)

Popkin, Jeremy D., *News and Politics in the Age of Revolution: Jean Luzac's Gazette de Leyde* (Cornell: Cornell University Press, 1989)

Portillo Valdés, José María, *Revolución de nación. Orígenes de la cultura constitucional en España* (Madrid: Boletín Oficial del Estado, 2000)

———, "Crisis e independencias: España y su monarquía," *Cuadernos dieciochistas* 8 (2007), 19–35

Proietti, Domenico, "La frammentazione dialettale e la situazione linguistico-culturale italiana nell'opera di Lorenzo Hervas y Panduro," in *La presenza dei gesuiti iberici espulsi. Aspetti religiosi, politici,* culturali, eds. Ugo Baldini and Gian Paolo Brizzi (Bologna: Clueb, 2010), 587–608

Pucci, Luigi, *Lodovico Ricci. Dall'arte del buon governo alla finanza moderna 1742–1799* (Milan: Giuffrè 1971)

Rauschning, Dietrich, *Georg Friedrich von Martens (1756–1821). Lehrer der praktischen Europäischen Völkerrechts und der Diplomatie zu Göttingen,* in *Rechtswissenschaft in Göttingen. Göttinger Juristen aus 250 Jahren,* ed. Fritz Loos (Göttingen: Vandenhoeck und Ruprecht, 1987), 123–145

Reinert, Sophus A., *The Academy of Fisticuffs: Political Economy and Commercial Society in Enlightenment Italy* (Cambridge Mass.: Harvard University Press, 2019).

———, *Translating Empire. Emulation and the Origins of Political Economy* (Cambridge, Mass.: Harvard University Press, 2011)

Remec, Peter Paul, *The Position of the Individual in International Law according to Grotius and Vattel* (The Hague: Martinus Nijhoff, 1960)

Repetti, Francesco, "Attività editoriale a Livorno tra Settecento e Ottocento: la stamperia di Tommaso Masi," *Nuovi Studi livornesi* 3 (1995), 92–125

Rigatti, Maria, *Un illuminista trentino del secolo XVIII: Carlo Antonio Pilati* (Florence: Vallecchi, 1923)

Robertson, John, *The Case for Enlightenment: Scotland and Naples 1680–1760* (Cambridge: Cambridge University Press, 2007)

Romano, Andrea, "Cadice come modello costituzionale per l'Europa liberale e antinapoleonica. Nota introduttiva," in *Costituzione politica della monarchia spagnola (1813),* ed. Andrea Romano (Soveria Monnelli: Rubbettino 2000), XVII–LXXXV

———, "L'influenza della carta gaditana nel costituzionalismo italiano ed europeo," in *La Constitucion de Cádiz de 1812: hacia los orígenes del constitucionalismo iberoamericano y latino,* ed. Asdrúbal Aguilar Aranguren (Caracas: Universidad Catolica Andreas Bello, 2004), 351–373

Rosa, Mario, *Settecento religioso. Politica della ragione e religione del cuore* (Venice: Marsilio, 1999)

Rosanvallon, Pierre, *Le bon gouvernement* (Paris: Editions du Seuil, 2016)

Roulet, Louis-Edouard, "Friedrich der Große und Neuenburg," in *Friedrich der Große in seiner Zeit*, ed. Oswald Hauser (Cologne-Vienna: Böhlau, 1987), 181–191

Roux, Christine, *Les 'Makis' de la resistance corse 1772–1778* (Paris: France Empire, 1984)

Ruata, Ada, *Luigi Malabaila di Canale. Riflessi della cultura illuministica in un diplomatico piemontese* (Turin: Deputazione subalpina di storia patria, 1968)

Sabbatini, Renzo, and Volpini, Paola (eds.), *Sulla diplomazia in età moderna. Politica, economia, religione* (Milan: FrancoAngeli, 2011)

Sabetti, Filippo, *The Search of Good Government. Understanding the Paradox of Italian Democracy* (Montreal: McGill-Queen's University Press, 2000)

——, (ed.), *Alla ricerca del buon governo in Italia* (Manduria: Lacaita, 2004)

Scarabello, Giovanni, "Il Settecento," in *La repubblica di Venezia in età moderna, dal 1517 alla fine della Repubblica*, vol. 2, eds. Gustavo Cozzi, Michael Knapton and Giovanni Scarabello (Turin: Utet, 1992), 551–681

Schilling, Lothar, *Kaunitz und das Renversement des alliances. Studien zur aussenpolitischen Konzeption Wenzel Anton Kaunitz* (Berlin: Duncker und Humblot, 1994)

Schlup, Michel, ed., *L'édition neuchâteloise au siècle des Lumières. La Société typographique de Neuchâtel (1769–1789)* (Neuchâtel: Bibliothèque Publique et Universitaire, 2002)

Schmidt-Voges, Inken, "Making Peace in Early Modern Europe," in *Peace was made here: the treaties of Utrecht, Rastatt and Baden 1713–1714*, eds. Renger de Bruin and Maarten Brinkman (Petersberg: Im Hof, 2013), 49–59

Schnakenborug, Eric, *Entre la guerre et la paix. Neutralité et relations internationales XVIIe-XVIIIe siècles* (Rennes: Presses Universitaires de Rennes, 2013),

Schnettger, Matthias, "*Kleinstaaten in der Frühen Neuzeit. Konturen eines Forschungsfeldes*," Historische Zeitschrift 236 (2008), 605–640

Schroeder, Francesco, *Repertorio biografico delle famiglie confermate nobili e dei titolati* (Venice: Alvisopoli, 1830)

Schroeder, Paul W., "Did the Vienna Settlement Rest on a Balance of Power?", *The American Historical Review*, 97 (1992), 683–706

Scina, Domenico, *Prospetto della storia letteraria di Sicilia nel secolo decimottavo*, vol. 1 (Palermo: Lo Bianco, 1859)

Scovazzi, Tullio, "The Evolution of International Law of the Sea: New Issues, New Challenges," in *Recueil des cours. Collected courses of the Hague Academy of International Law*, ed. Académie de droit international (The Hague: Nijhoff, 2001), 39–244

Silvestri, Paolo, *Il liberalismo di Luigi Einaudi, o del Buongoverno* (Soveria Mannelli: Rubbettino, 2008)

Skinner, Quentin, "Ambrogio Lorenzetti: The Artist as Political Philosopher," *Proceedings of the British Academy* 72 (1986), 1–6

———, "Ambrogio Lorenzetti's buon governo frescoes: Two old questions, two new answers," *Journal of the Warburg and Courtauld Institutes* 62 (1998), 1–8

———, "Il Buon governo di Ambrogio Lorenzetti e la teoria dell'autogoverno repubblicano," in Adorni Braccesi, Simonetta, and Ascheri, Mario (eds.), *Politica e cultura nelle Repubbliche italiane dal Medioevo all'età moderna. Firenze, Genova, Lucca, Siena, Venezia* (Rome: Istituto storico italiano per l'età moderna e contemporanea, 2001), 21–42

———, *The Foundation of Modern Political Thought*, vol. 1, *The Renaissance* (Cambridge: Cambridge University Press, 1978)

———, *Visions of Politics*, vol. 2, *Renaissance Virtues* (Cambridge: Cambridge University Press, 2002)

Sordi, Bernardo, *L'amministrazione illuminata. Riforma delle comunità e progetti di costituzione nella Toscana leopoldina* (Milan: Giuffrè, 1991)

Spaggiari, Wlliam, *L'armonico tremore. Cultura settentrionale dall'Arcadia all'età napoleonica* (Milan: FrancoAngeli, 1990)

Spini, Giorgio, *Mito e realtà della Spagna nelle rivoluzioni italiane del 1820–1821* (Rome: Perrella, 1950)

Stapelbroek, Koen, "'The long peace': commercial treaties and the principles of global trade at the Peace of Utrecht," in: *The 1713 Peace of Utrecht and its enduring effects*, ed. A.H.A. Soons (Leiden: Brill, 2019a), 93–119

———, "L'organisation du commerce international dans l'ombre d'Utrecht: les perspectives hollandaises du XVIIIe siècle," in *La paix d'Utrecht (1713): Enjeux économiques, maritimes et commerciaux*, eds. Lucien Bély, Géraud Poumarède and Guillaume Hanotin (Paris: Pedone, 2018), 475–502

———, "The Foundations of Vattel's "System" of Politics and the Context of the Seven Year's War: Moral Philosophy, Luxury and the Constitutional Commercial State," in *The Legacy of Vattel's Droit des gens: Contexts, Concepts, Reception, Translation and Diffusion*, eds. Koen Stapelbroek and Antonio Trampus (Cham: Palgrave Macmillan, 2019b, 95–123

———, "Universal Society, Commerce and the Rights of Neutral Trade: Martin Hübner, Emer de Vattel and Ferdinando Galiani," in: *Universalism in International Law and Political Philosophy*, ed. Petter Korkman and Virpi Mäkinen, *Collegium. Studies across Disciplines in the Humanities and Social Sciences* 4 (2008a) (Helsinki: Helsinki Collegium for Advanced Studies, 2008), 63–89

———, *Love, Self-deceit and Money* (Toronto: University of Toronto Press, 2008b)

———, and Trampus, Antonio, "*Commercial reform against the tide: Reapproaching the eighteenth-century decline of the republics of Venice and the United Provinces*" in: *History of European Ideas*, vol. 36 (2010), 192–202

————, and Trampus, Antonio, "The Legacy of Vattel's Droit des gens: Contexts, Concepts, Reception, Translation and Diffusion," in: in *The Legacy of Vattel's Droit des gens: Contexts, Concepts, Reception, Translation and Diffusion*, eds. Koen Stapelbroek and Antonio Trampus (Cham: Palgrave Macmillan, 2019), 1–25

Stolleis, Michael, *Public Law in Germany: A Historical Introduction from the 16th to the 21st Century* (Oxford: Oxford University Press, 2017)

Tabacchi, Stefano, *Il Buon governo: le finanze locali nello Stato della Chiesa (secoli XVI–XVIII)* (Rome: Viella, 2007)

Tabacco, Giovani, *Andrea Tron e la crisi dell'aristocrazia senatoria a Venezia* (Udine: Del Bianco, 1983)

Tavoni, Maria Grazia, "Tipografi, editori, lettura," in *Storia di Bologna*, eds. Aldo Berselli and Angelo Varni, vol. 4/1 (Bologna: Bononia University Press, 2010), 687–868

Tolomio, Ilario, *Theism and the History of Philosophy: Appiano Buonafede*, in *Models of the History of Philosophy, III. The Second Enlightenment and the Kantian Age* (New York: Springer, 2015), 359–382

Tomás y Valiente, Francisco, "Génesis de la Constitution de 1812," *Anuario de Historia del Derecho español* LXV (1995): 13–125

————, *Génesis de la Constitución de 1812. De muchas leyes fundamentales a una sola constitución* (Pamplona: Urgoiti, 2012)

Torcellan, Gianfranco, "Cesare Beccaria e Venezia," *Rivista Storica Italiana* LXXVI (1964), 720–748

Toyoda, Tetsuya, "Vattel's doctrine of national sovereignty in the context of Saxony Poland and Neuchâtel," in *Theory and politics on the law of nations: Political Bias in International Law Discourse of Seven German Court Councilors in the Seventeenth and Eighteenth Centuries* (Leiden: Brill, 2011), 161–190

Trampus, Antonio, "Enlightenment in Global History: On Filangieri's Science of Legislation and the Transformation of Political Language in the Classical Liberalism", in Век Просвещения. Что такое Просвещение? Новые ответы на старый вопрос *(Le Siècle des Lumières. Qu 'est-ce que les Lumières? Nouvelles reponses à l'ancienne question)*, ed. Serguei Karp et al. (Moscow: Nauka, 2019), 110–125

————, "Il ruolo del traduttore nel tardo Illuminismo: Lodovico Antonio Loschi e la traduzione italiana del *Droit des gens*," in *Il linguaggio del tardo Illuminismo*, ed. Antonio Trampus (Rome: Edizioni di Storia e Letteratura, 2009), 81–109

————, "La genesi e la circolazione della 'Scienza della legislazione'. Saggio bibliografico," *Rivista Storica Italiana* CXVII (2005), 309–359

————, "The circulation of Vattel's Droit des gens in Italy: the doctrinal and practical model of government," in *War, Trade and Neutrality: Europe and the Mediterranean in the seventeenth and eighteenth centuries*, ed. Antonella Alimento (Milan: FrancoAngeli, 2011), 217–232

————, "Tra Corsica e Toscana: Emer de Vattel e i percorsi del costituzionalismo settecentesco," *Etudes Corses* 78 (2014), 61–80

————, *Storia del costituzionalismo italiano nell'età dei Lumi* (Rome-Bari: Laterza, 2009)

Trentafonte, Franco, *Giurisdizionalismo, Illuminismo e massoneria nel tramonto della Repubblica veneta* (Venice: Deputazione editrice, 1984)

Tribe, Keith, *Governing Economy: The Reformation of German Economic Discourse 1750–1840* (Cambridge: Cambridge University Press, 1988)

Tuck, Richard., *The Rights of War and Peace: Political Thought and the International Order From Grotius to Kant* (Oxford: Oxford University Press: 1999)

Tufano, Roberto, "Il popolo nel governo di Bernardo Tanucci. L'emergenza della questione sociale nel Regno di Napoli (1734–1774)," in *Un'isola nel contesto mediterraneo. Politica, cultura e arte nella Sicilia e nell'Italia meridionale in età medievale e moderna*, eds. Carmelina Urso, Paola Vitolo and Emanuele Piazza (Bari: Adda, 2018), 103–148

Turi, Gariele, *Brichieri Colombi Giovanni Domenico*, in *Dizionario biografico degli italiani*, vol. 14 (Rome: Istituto della Enciclopedia italiana, 1972), 229–232

Valensise, Marina, "La constitution française," in *The French Revolution and the Creation of Modern Political Culture*, vol. 1, *The Political Culture of the Old Regime*, ed. Keith M. Baker (Oxford: Oxford University Press, 1987), 445–446

Valera, Valera, *Scienza dello Stato e metodo storiografico nella scuola storica di Gottinga* (Naples: Edizioni Scientifiche Italiane, 1980)

Vannini, Fabrizio, "Lampredi Giovanni Maria," in *Dizionario biografico degli italiani*, vol. 63 (Rome: Istituto della Enciclopedia Italiana, 2004), 259–262

Venturi, Franco, "Economisti e riformatori spagnoli e italiani del '700," *Rivista storica italiana* LXXIV (1962a), 532–561

————, "Ritratto di Agostino Paradisi," *Rivista Storica Italiana* 74 (1962b), 717–738

————, *Illuministi italiani 3, Riformatori lombardi, piemontesi e toscani* (Milan-Naples: Ricciardi, 1958)

————, *Settecento riformatore* vol. 2, *La Chiesa e la Repubblica dentro i loro limiti (1758–1785)* (Turin: Einaudi, 1976),

————, *Settecento riformatore*, vol. 5/1, *L'Italia dei Lumi (1764–1790)*, (Turin: Einaudi, 1987)

————, *Settecento riformatore*, vol. 5/2, *L'Italia dei Lumi (1764–1790). La Repubblica di Venezia (1761–1797)*, (Turin: Einaudi, 1990)

————, *Settecento riformatore* vol. 4/2 *La caduta dell'Antico Regime 1776–1789* (Turin: Einaudi, 1984)

————, *Venezia nel secondo Settecento* (Turin: Tirrenia Stampatori, 1980)

Verga, Marcello, *Da "cittadini" a "nobili". Lotta politica e riforma delle istituzioni nella Toscana di Francesco Stefano* (Milan: Giuffrè, 1990)

Vick, Brian E., *The Congress of Vienna. Power and Politics after Napoleon* (Cambridge, MA: Harvard University Press, 2014)

Vitelli, Tom, "Casanova and Gassendi. Proposition for a study," *Casanova Gleanings* 1980, 11–14

Viviani Della Rocca, Enrica, *Bernardo Tanucci e il suo più importante epistolario* vol. 2 (Florence: Sansoni, 1942)

Von Laue, Theodore H., *Leopold von Ranke. The Formative Years* (Princeton: Princeton University Press, 1950)

Wandruska, Adam, *Leopold II* (Vienna: Herold, 1963; it. translation: *Pietro Leopoldo. Un grande riformatore*, Florence: Vallecchi, 1968)

———, "Joseph II und das Verfassungsproject Leopolds II," *Historische Zeitschrift* 190 (1960), 18–30

Whatmore, Richard, *Against War and Britain, and France in the Eighteenth Century* (Yale: Yale University Press, 2012)

———, and Kapossy, Béla, "Emer de Vattel's Mélanges de littérature, de morale et de politique (1760)," *History of European Ideas* 34 (2008), 77–103

Zannoni, Giovanni, *Una lettera inedita di Carlo Innocenzo frugoni a Lodovico Antonio Loschi* (Rome: Tip. Elzeviriana, 1895)

Zapperi, Roberto, "Edmund Burke in Italia," *Cahier Vilfredo Pareto* 7–8 (1965), 38–40

Zimmermann, Joseph, *Das Verfassungsproject des Grossherzogs Peter Leopold von Toskana* (Heidelberg: Carl Winter, 1901)

Zordan, Giorgio, *Il Codice per la Veneta mercantile marina* (Padua: Cedam, 1987)

Zurbuchen, Simone, "Das Verhältnis Europas zu den Staaten der Alten und der Neuen Welt. Die Idee einer société générale du genre humain in Emer von Vattels Völkerrecht," in *Europa und die Moderne im langen 18. Jahrhundert*, ed. Olaf Asbach (Hanover: Wehrhahn Verlag, 2014a), 167–188

———, "Die Schweiz, in Grundriss der Geschichte der Philosophie. Die Philosophie des 18. Jahrhunderts," *Heiliges Römisches Reich Deutscher Nation. Schweiz, Nord- und Osteuropa*, vol. 5, ed. Helmut Holzhey and Vilem Mudroch (Basel: Schwabe, 2014b), 1445–1485

———, "Vattel's law of nations and just war theory," *History of European Ideas* 35 (2009), 408–417

Index[1]

[1] Note: Page numbers followed by 'n' refer to notes.

© The Author(s) 2020
A. Trampus, *Emer de Vattel and the Politics of Good Government*,
https://doi.org/10.1007/978-3-030-48024-0

Printed in the United States
by Baker & Taylor Publisher Services